Non-Linear Spectral Unmixing of Hyperspectral Data

This book is based on satellite image processing, focusing on the potential of hyperspectral image processing (HIP) research with a case study-based approach. It covers the background, objectives, and practical issues related to HIP and substantiates the needs and potentials of said technology for discrimination of pure and mixed endmembers in pixels, including unsupervised target detection algorithms for extraction of unknown spectra of pure pixels. It includes application of machine learning and deep learning models on hyperspectral data and its role in spatial big data analytics.

Features include the following:

- Focuses on capability of hyperspectral data in characterization of linear and non-linear interactions of a natural forest biome.
- Illustrates modeling the ecodynamics of mangrove habitats in the coastal ecosystem.
- Discusses adoption of appropriate technique for handling spatial data (with coarse resolution).
- Covers machine learning and deep learning models for classification.
- Implements non-linear spectral unmixing for identifying fractional abundance of diverse mangrove species of coastal Sundarbans.

This book is aimed at researchers and graduate students in digital image processing, big data, and spatial informatics.

Non-Linear Spectral Unmixing of Hyperspectral Data

Somdatta Chakravortty

CRC Press
Taylor & Francis Group
Boca Raton London New York

CRC Press is an imprint of the
Taylor & Francis Group, an **informa** business

Designed cover image: image by vecstock on Freepik

First edition published 2025
by CRC Press
2385 NW Executive Center Drive, Suite 320, Boca Raton FL 33431

and by CRC Press
4 Park Square, Milton Park, Abingdon, Oxon, OX14 4RN

CRC Press is an imprint of Taylor & Francis Group, LLC

ISBN: 9781032450490 (hbk)
ISBN: 9781032558615 (pbk)
ISBN: 9781003432623 (ebk)

DOI: 10.1201/9781003432623

Typeset in Times
by Newgen Publishing UK

Contents

Preface

Welcome to the world of hyperspectral data processing and analysis. In this book we delve into the fascinating field of hyperspectral remote sensing and its applications. With advancements in technology, the availability of hyperspectral data has opened up new possibilities for understanding our environment and extracting valuable information from complex data sets.

Hyperspectral imaging offers a unique perspective, allowing us to capture and analyze a vast range of spectral information for each pixel in an image. This wealth of data presents both challenges and opportunities. The processing and analysis of hyperspectral data require specialized techniques and algorithms to unravel the rich information embedded within the spectra.

Our aim with this book is to provide a comprehensive overview of the key concepts, methodologies, and practical applications related to hyperspectral data processing. We cover a wide range of topics, from preprocessing techniques to endmember detection, unmixing algorithms, and machine learning models. Each chapter is meticulously crafted to provide a deep understanding of the subject matter while maintaining a practical approach.

The objective of this book is to provide a comprehensive review of hyperspectral image processing and its wide-ranging applications. It begins by establishing a solid foundation, offering a thorough understanding of the field's background, challenges, and potential. We discuss the limitations of earlier surveys based on multispectral imagery and emphasize the need for hyperspectral technology to discriminate pure and mixed endmembers accurately in pixels.

To address the challenges posed by hyperspectral data, we explore unsupervised endmember detection algorithms and analyze their results. We delve into linear and non-linear spectral unmixing techniques, aiming to achieve improved and accurate estimation of mixed pixel abundances. Additionally, we delve into the application of machine learning and deep learning models for the classification of hyperspectral data, enabling the extraction of meaningful information from complex imagery.

Moreover, this book goes beyond traditional applications and delves into the dynamics of coastal forest ecosystems. Through ecodynamic models based on time-series analyses of hyperspectral image data at sub-pixel levels, we uncover insights into the intricate relationships and competition among species. This knowledge aids in understanding the stability, dominance, and mutual exclusion of mangrove endmembers over time, contributing to the conservation and management of these critical habitats.

Throughout the book, we present a case study focused on the Sundarban mangroves in West Bengal, India. We navigate the complexities of extracting and modeling spectral information from mixed pixels, showcasing the effectiveness of non-linear and higher-order models in accurately classifying different mangrove species. To provide practical insights, we present a case study focused on the species-level discrimination of mangroves using hyperspectral image data. The study highlights the importance of non-linear spectral unmixing and its application in estimating the fractional abundance of mangrove species. We showcase the efficacy of higher-order models in characterizing the complex interactions between different endmembers, overcoming the limitations of previous linear and bilinear models.

It is my hope that this book serves as a valuable resource for researchers, scientists, students, and practitioners who wish to delve into the fascinating world of hyperspectral image processing. The knowledge and insights presented here will enable readers to harness the potential of hyperspectral imagery for a wide range of applications, contributing to sustainable land management, resource exploration, efficient agriculture practices, responsible forestry, and enhanced national security.

I extend my gratitude to all the contributors, researchers, and experts who have shared their knowledge and expertise to make this book a reality. I would specifically like to thank my PhD students – Ms. Dipanwita Ghosh, Mr. Prem Kumar, and Ms. Srirupa Das –who have helped in implementing the models used in the book. I also express our appreciation to the readers, whose curiosity and enthusiasm fuel the advancement of remote sensing and its applications.

In conclusion, this book serves as a comprehensive guide for researchers, professionals, and students interested in hyperspectral remote sensing. My hope is that it will inspire readers to explore the vast potential of hyperspectral data, drive further innovation in this field, and contribute to the sustainable management of our natural resources.

Enjoy your journey into the realm of hyperspectral data processing and analysis!

About the Author

 **Somdatta Chakravortty** is an accomplished academician and researcher, currently serving as an Associate Professor and Head of the Department of Information Technology at Maulana Abul Kalam Azad University of Technology in West Bengal, India. With a diverse educational background, she holds a B.Tech. degree from HBTI, Kanpur University, India an M.Tech. degree from Bengal Engineering and Science University, Sibpur, India and a Ph.D. degree from Calcutta University, Kolkata, India.

Dr. Chakravortty's professional journey encompasses a range of teaching and industry experiences. She has worked as in Consulting Engineering Services (India) Limited, Kolkata, and later served at various institutes, including Dr. B.C. Roy Engineering College, Durgapur, MCKV Institute of Engineering, Howrah and Heritage Institute of Technology, Kolkata. From 2008 to 2018, she held the position of Assistant Professor in Information Technology at Govt. College of Engineering & Ceramic Technology, Kolkata.

Her research interests have been instrumental in securing major and minor research projects funded by esteemed central government organizations such as the Department of Science and Technology, University Grants Commission, and All India Council of Technical Education. Dr. Chakravortty has made significant contributions to the field, with publications in renowned national and international journals and conferences and actively contributes as a reviewer for esteemed journals. She also has copyright on her work on spectral indices and has applied for patent on her work on hyperspectral spectra generation.

As a passionate educator and mentor, Dr. Chakravortty has guided several Ph.D. students and supervised many M.Tech dissertations. Her current research focuses on emerging areas in the field of image processing, hyperspectral remote sensing, image fusion, and machine learning.

Dr. Chakravortty actively engages with professional societies such as the IEEE, Indian Society of Remote Sensing, the Computer Society of India, the Institution of Engineers, and the Association of Engineers.

1 Introduction

1.1 HYPERSPECTRAL IMAGE PROCESSING

Hyperspectral image processing is a powerful tool for extracting and analyzing spectral information from images. It has various applications in fields such as remote sensing, mineral exploration, agriculture, forestry, and military surveillance. In remote sensing, hyperspectral images can be used to identify and map different types of vegetation, minerals, and land cover. In mineral exploration, hyperspectral images can be used to identify and map mineral deposits. In agriculture, hyperspectral images can be used to monitor crop growth, detect pests, and assess soil conditions. In forestry, hyperspectral images can be used to map tree species and identify areas of disease or damage. In military surveillance, hyperspectral images can be used to detect camouflage and other military targets.

However, hyperspectral image processing has its challenges. One of the main challenges is the high dimensionality of the data, making it difficult to extract relevant information and classify different materials or objects in the scene. To address this challenge, dimensionality reduction techniques such as principal component analysis (PCA) or independent component analysis (ICA) are often used to reduce the number of spectral bands and improve computational efficiency. Additionally, feature extraction techniques such as the use of spectral indices can be used to extract relevant information from the data.

Another challenge is noise and artifacts in the data, making it challenging to identify and classify different materials or objects in the scene. To address this challenge, data preprocessing techniques such as atmospheric correction and noise reduction are used to remove artifacts and improve the quality of the images.

Despite these challenges, hyperspectral image processing has the potential to revolutionize a wide range of fields by providing detailed information about the spectral characteristics of a scene. With advancements in sensor technology, data processing algorithms, and machine learning, the capabilities of hyperspectral image processing will continue to expand and become more powerful in the future.

DOI: 10.1201/9781003432623-1

1.2 AVAILABILITY OF HYPERSPECTRAL DATA

Hyperspectral data is available from various sources, including government agencies, commercial providers, research institutions, and open-source data sets. The data can be obtained from commercial and government satellites, as well as handheld and airborne systems, depending on the application and the area of interest.

Government agencies such as NASA and the US Geological Survey (USGS) collect and make hyperspectral data open to the public. These data are often collected using satellite or aircraft-based sensors and cover various locations and applications. NASA's Hyperion sensor and the USGS's AVIRIS sensor are examples of hyperspectral sensors used to collect data for multiple projects. The Hyperion sensor, for example, covers a wavelength range from 0.4 to 2.5 micrometers and a spatial resolution of 30 meters. The AVIRIS sensor covers a wavelength range from 0.4 to 2.5 micrometers and has a spatial resolution of 20 meters.

Commercial providers such as Digital Globe and Airbus Defence and Space offer hyperspectral data for a fee. These data may be collected using satellite or aircraft-based sensors or through partnerships with government agencies. Commercial providers often offer higher-resolution data and faster access times than government agencies, but the cost may be prohibitive for some applications. For example, the WorldView-3 satellite from Digital Globe can provide data with a resolution of 0.3 meters. In comparison, Airbus's Hyperspectral Imager for the Coastal Ocean (HICO) sensor can provide data with a resolution of 30 meters.

Research institutions such as the University of California, Santa Barbara, and the Technical University of Denmark also collect and make hyperspectral data available to the public. These data may be collected as part of a specific research project or made available through partnerships with government agencies or commercial providers. Research institutions often use specialized sensors and equipment, such as the Airborne Visible/Infrared Imaging Spectrometer (AVIRIS) sensor or the Reflective Optics System Imaging Spectrometer (ROSIS) sensor, to collect data for specific research projects.

Open-source data sets, such as the hyperspectral data collected by the AVIRIS sensor, are also available online. These data sets can be helpful for researchers and students interested in hyperspectral image processing. Open data platforms such as the NASA Earth Data portal and the European Space Agency's Sentinel Data Hub provide access to a wide range of hyperspectral data, covering different wavelength ranges, spatial and spectral resolutions, and acquisition conditions.

It is also possible to collect hyperspectral data using specialized sensors and equipment. However, this process can be complex and costly, requiring specialized training and expertise. Before using a specific dataset, it is essential to carefully evaluate its suitability for specific needs and goals, considering factors such as the wavelength range, resolution, and acquisition conditions.

1.3 THE OBJECTIVE OF THE BOOK

The objective of the book covers the following:

(i) Review of hyperspectral image processing and its applications.
(ii) Exploring unsupervised endmember detection algorithms on hyperspectral data and result analysis.
(iii) Exploring and implementing linear and non-linear spectral unmixing of higher-order for improved and accurate estimation of mixed pixel abundances.
(iv) Application of machine learning and deep learning models for classification of hyperspectral data.
(v) Understanding the dynamics of coastal forest ecosystem through ecodynamic models based on time-series analyses of hyperspectral image data at sub-pixel levels.

1.4 CONTENT AND ORGANIZATION OF THE BOOK

The initial chapters of the book detail the study background, objectives, and moot issues related to hyperspectral image processing. It substantiates the needs and potential of hyperspectral technology for the discrimination of pure and mixed endmembers in pixels, which is the main objective of this book. The limitations of the earlier surveys based on multispectral imagery have been brought out vividly in the initial chapters.

The chapters undertake critical analysis and appreciation of the unsupervised target detection algorithms on hyperspectral data as a probable solution for extracting unknown spectra of pure pixels from the hyperspectral image.

Least-squares-based linear spectral unmixing for pure endmembers for analysis of linearly arranged and homogeneous endmember groups enunciates the applications of spectral signatures of pure endmembers extracted through automated target detection algorithms. The derived spectral information is then transferred to achieve the fractional abundance values through linear spectral unmixing model. This model helps to identify the location of endmembers and correlate them with the actual physical abundance of endmembers on the particular image scene.

Non-linear unmixing for classification of mixed endmembers elaborates the non-linear model that seeks to identify the mixed endmembers by considering higher-order interactions (up to the 'nth order') between the different endmembers and shows how best they validate the actual ground conditions. This was so far neglected in the linear spectral unmixing model. The non-linear models incorporate the interaction terms between similar endmembers to better represent fractional abundances. This model has been applied to the studied habitats of Sunderban mangroves, and the application proved highly successful.

The book also elaborates on applying machine learning and deep learning models to hyperspectral data and its role in spatial big data analytics.

Modeling the ecodynamics of coastal forest areas has successfully made use of two famous ecological models, namely the ecodynamic model and the Lotka–Volterra competition model, in explaining the inherent intra- and inter-species competition of natural forests over a time period. The model is based on image-derived fractional abundances obtained after linear and non-linear spectral unmixing. This has helped to determine the dominance of mangrove endmembers in a pixel area and examine its state of stability. The mangrove endmember stability and dynamics within mixed patches have been successfully brought out to predict species coexistence, dominance, and mutual exclusion over a period of time.

1.5 SOCIETAL RELEVANCE

Hyperspectral image processing is increasingly necessary in addressing a wide range of societal challenges. The technology is used in many fields, including remote sensing, mineral exploration, agriculture, forestry, and military surveillance, which are all critical for economic and environmental sustainability.

In remote sensing, hyperspectral imaging is used to monitor land use, vegetation, mineral resources, and natural disasters. This information is used to support sustainable land management, natural resource conservation, and disaster response efforts. For example, hyperspectral data can detect and map vegetation changes, identify land degradation areas, and monitor forests' health, which is critical for maintaining biodiversity and combating climate change.

In mineral exploration, hyperspectral imaging identifies and maps mineral resources, including minerals, oil, and gas. This information supports sustainable and responsible resource extraction, which is critical for economic growth and development.

In agriculture, hyperspectral imaging is used to monitor crop health and productivity, identify areas of soil erosion and nutrient deficiencies, and detect and map pests and diseases. This information supports sustainable and efficient food production, which is critical for food security and reducing poverty.

In forestry, hyperspectral imaging is used to map tree species, monitor forest health, and detect areas of illegal logging or deforestation. This information is used to support sustainable forest management, which is critical for maintaining biodiversity and combating climate change.

In military surveillance, hyperspectral imaging is used to detect and map changes in land use, vegetation, and mineral resources and detect and track the movement of people and vehicles. This information is used to support national security and public safety.

Overall, hyperspectral image processing is a powerful tool for extracting and analyzing spectral information from images and plays a critical role in addressing a wide range of societal challenges.

1.6 A CASE STUDY OF MANGROVE ENDMEMBERS FOR RESULT ANALYSIS

This book aims to extract the potential of hyperspectral imagery in forest classification and the consequent forest cover mapping. The case study is carried out in the biodiverse rainforest of Sunderban in the State of West Bengal, India. The biologically diverse Sunderban mangroves in close clusters are ideal exemplars for analyzing non-linear scattering. The interactions thus envisaged between adjacent mangrove species of varying biophysical manifestations produce non-linearities in the system. The diverse spectral information embedded within such mixed pixels is a big challenge for extraction and needs to be modeled effectively for the desired outcome.

Mangrove habitats are typically marked by the dominance of image pixels of vegetation, soil, and water systems. The surface features of mangrove communities are greatly influenced by the inter-tidal environment that gets reflected in the spectral signatures of the image pixels. Furthermore, the species diversity of Sunderban mangroves is much higher than any other tropical mangrove forest in the world (Jensen, 1996). Such wide pattern recognition poses an uphill task for RS-SIP analyses, as increasing diversity leads to discrimination problems due to the presence of innumerable spectrally unique species.

The mangroves of Henry Island of Sunderban (the study area) are dominated by *Excoecaria agallocha*, *Ceriops decandra*, *Avicennia marina*, and *Avicennia alba*, which constitutes about 70% of the mangrove species in the Indian Sunderbans. Out of the 94 species of mangroves reported from the Sunderban, only seven pure (homogeneous) patches of species have been traced on this island. The habitats of a wide range of other mangrove species of Sunderban are practically inaccessible to man. The mangroves of Henry are clustered in zones with a dominance of the species of *Avicennia* and *Excoecaria* on the seaward edge. Away from the water line, lie the mixed patches of *Bruguiera cylindrica*, *Phoenix paludosa*, *Ceriops decandra*, and *Excoecaria agallocha*. *Sundari*, the climax mangrove species of the Sunderban, is in decline in Henry owing to high salinity hazard. A pure patch of *Sundari* in a 30m × 30m matrix is virtually absent here. Although artificial afforestation programs of *Sundari* have been taken up by the Forest Department at select patches, the efforts are far from satisfactory. The dense mangrove forest of Henry is crisscrossed by numerous small canals with evidence of wildlife, namely deer, wild pig, crocodile, and poisonous snakes, posing threats to man.

The study area offers an ideal locale for the application of hyperspectral remote sensing for mapping and monitoring a wide range of mangrove species of pure and mixed patches (Chowdhury and Maiti, 2014; Samanta and Hazra, 2012). The ecological fragility of the Sunderban rainforest (Nandy and Kushwaha, 2010; Singh et al., 2006) is reflected by the dwindling density and diversity of the mangroves of Henry. This can be successfully interpreted through the study of different endmembers (mangrove species) that cast a profound influence on the integrated whole.

The typical image pertaining to a particular hyperspectral band of the study area (Henry Island) is shown in Figure 1.1.

This book discusses the application of non-linear and higher-order models for the classification of mixed endmembers in a hyperspectral dataset of mangrove species. The dataset, collected from Henry Island in the Sunderbans, has been analyzed using a variety of methods, including Nascimento's bilinear spectral unmixing model, Fan's bilinear unmixing model, and higher-order models.

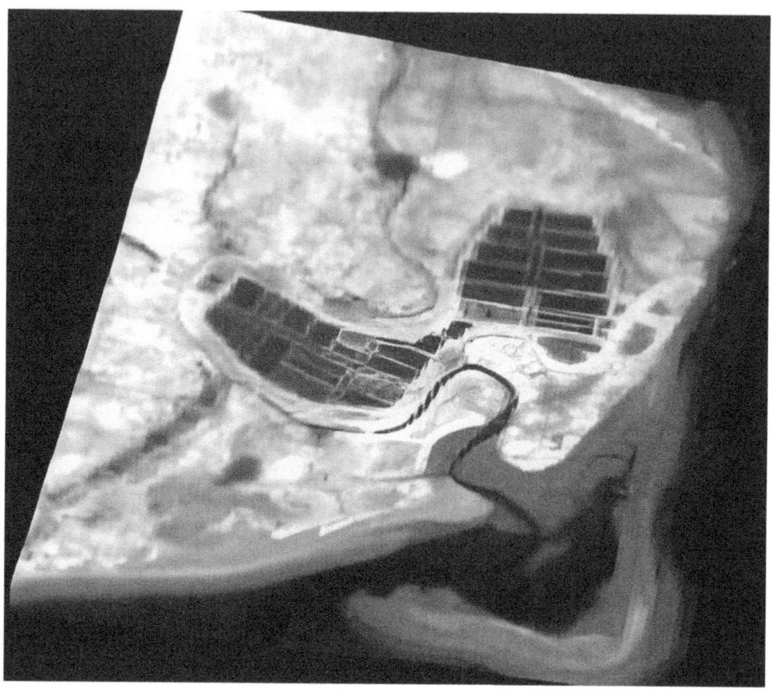

FIGURE 1.1   Hyperspectral imagery of Henry Island.

The application of these methods on a dataset of mangrove species has shown their ability to accurately classify the different species present in the dataset. Overall, this case study highlights the importance of using non-linear and higher-order methods in the analysis of hyperspectral data and the potential for these methods to aid in the identification and management of critical ecological systems such as mangroves.

2 Hyperspectral Image Processing

A Review

2.1 EXISTING MULTISPECTRAL TECHNOLOGY AND ITS LIMITATIONS

Existing multispectral technology, such as multispectral imaging sensors and cameras, capture reflectance information at a limited number of discrete wavelength bands, typically a few to several dozen. These systems are commonly used in remote sensing, mineral exploration, agriculture, and forestry.

While multispectral technology has been successful in many applications, it has several limitations when compared to hyperspectral technology. One of the main limitations is the limited number of wavelength bands, which results in less detailed and accurate spectral information. This can make it more difficult to accurately identify and classify materials or objects in the scene.

Another limitation of multispectral technology is that it is often less sensitive to subtle variations in reflectance, making it more difficult to detect and map small or subtle environmental changes. This can be particularly challenging in applications such as mineral exploration, where small changes in mineral composition can have a big impact on resource extraction.

Additionally, multispectral technology is often less precise in terms of spatial resolution, making it more difficult to accurately map and monitor environmental changes at a fine scale. This can be particularly challenging in applications such as agriculture, where precise mapping of crop health and productivity is critical for efficient food production.

In contrast, hyperspectral technology captures reflectance information at hundreds or thousands of narrow, contiguous wavelength bands, resulting in much more detailed and accurate spectral information. This makes it possible to accurately identify and classify materials or objects in the scene, even when they are present in small amounts or exhibit subtle variations in reflectance. Additionally, hyperspectral technology often has higher spatial resolution than multispectral technology, making it possible to accurately map and monitor environmental changes at a fine scale.

One of the advantages of hyperspectral technology is the ability to use advanced spectral unmixing algorithms that can decompose the mixed pixels of different materials or substances into the endmembers (pure pixels) that compose them. This

DOI: 10.1201/9781003432623-2

allows for the estimation of abundance fractions of each endmember, which can provide a more detailed understanding of the composition of the scene.

Another advantage of hyperspectral technology is the ability to use advanced machine learning algorithms for classification, which can be trained to recognize complex patterns and features in the data. These algorithms can provide more accurate and robust classifications than traditional methods, making it possible to identify even rare or subtle classes of materials or objects.

Overall, while multispectral technology has been successful in many applications, it has several limitations when compared to hyperspectral technology. The main advantages of hyperspectral technology include its ability to capture more detailed and accurate spectral information, its ability to use advanced spectral unmixing algorithms, and its ability to use advanced machine learning algorithms for classification.

This book focuses on mangrove species classification using hyperspectral images and how hyperspectral imaging can be used to improve the accuracy and precision of this task. In the case of the coastal mangrove ecosystem, traditional multispectral technology can have several limitations. One of the main limitations is the limited number of wavelength bands captured by multispectral sensors. This can make it difficult to accurately identify and classify different mangrove species, particularly when they have similar spectral characteristics. The limited number of wavelength bands also makes it harder to detect subtle variations in reflectance that may be important for distinguishing between different species.

While multispectral technology has been used for mangrove species classification in the past, it has several limitations that can affect the accuracy and precision of the results. Hyperspectral imaging, with its ability to capture more detailed and accurate spectral information, its ability to use advanced spectral unmixing algorithms, and its ability to use advanced machine learning algorithms, can help overcome these limitations and improve the accuracy and precision of mangrove species classification.

Several authors (Aschbacher et al., 1995; Jensen, 1996; Held et al., 2003; Saxena et al., 2004) had earlier studied in a broad way the spatial changes in the mangrove cover of Sunderban using IRS-P6 LISS III data. They employed a variety of Digital Image Processing techniques for enhancement and classification processes. It was observed that the middle infra-red band (0.77–0.86mm) is suitable for the mapping of mangroves. Two major classes of mangroves were delineated from the digital data.

Prasad et al. (1992) mapped and analyzed the vegetation, including mangrove types, of the North Andaman Islands (India) using LISS III satellite data. Singh et al. (2014) used Landsat-TM and IRS-P6 LISS-III data from 1987, 1998, 2000, and 2002 for broad-scale mangrove mapping and monitoring, database creation, and conservation prioritization of Sunderban mangroves, although in a limited way. To extract appropriate information from the imagery, different band combinations were used in these studies. It was reported that Landsat-TM bands 5, 6, 1 and LISS-III bands 3, 4, and 1 give satisfactory results for the delineation of mangroves dominated by Phoenix paludosa and Avicennia sp.

Zhi-Gang et al. (2008) worked on local spatial statistics for remotely sensed image classification of mangroves in Malaysia. It was revealed that as the spatial resolution improves in remotely sensed imagery, more detailed spatial information of mangrove forests could be shown. In this study, the SPOT-5 image of Matang Mangrove Forest Reserve (Malaysia) was used in mangrove forest classification. The results showed the need to find valuable features from high-resolution images to improve the mangrove classification accuracy.

Kanniah et al. (2007) applied linear spectral unmixing on IKONOS data to classify a mangrove forest in Malaysia and map three different mangrove species. The accuracy obtained through this procedure was found better than the minimum distance to mean classifier and the maximum likelihood classification with the inclusion of texture information.

From the above literature, it becomes clear that the broad spectral information of the sensors so far discussed can only be used in a limited way to characterize a few reflection and absorption spectral curves. It was revealed that the red edge, which represents the unique features of plant spectral responses between 690 nm and 720 nm wavelength, could hardly give any meaningful data under the existing provisions of multispectral sensors. Hence these sensors are of little use in the extraction of critical physio-chemical characteristics of plants, including the chlorophyll content (Goel, 2003; Elvidge, 1987; Himmelsbach, 1988; Curran, 1989; Kumar, 2001; Williams and Norris, 2001).

2.2 POTENTIALS OF HYPERSPECTRAL REMOTE SENSING

Hyperspectral image processing has the potential to significantly improve the accuracy and precision of various applications such as mineral exploration, agriculture, forestry, military surveillance, and mangrove species classification. Some of the critical potentials of hyperspectral image processing include the following:

- Detailed and accurate spectral information: hyperspectral image processing allows for the analysis of detailed and accurate spectral information captured at hundreds or thousands of narrow, contiguous wavelength bands. This makes it possible to accurately identify and classify different materials or objects in the scene, even when they have similar spectral characteristics or exhibit subtle variations in reflectance.
- Advanced spectral unmixing algorithms: hyperspectral image processing allows for the use of advanced spectral unmixing algorithms that can decompose mixed pixels into endmembers (pure pixels) and estimate the abundance fractions of each endmember. This provides a more detailed understanding of the composition of the scene and can help identify the presence of rare or subtle classes of materials or objects.
- Advanced machine learning algorithms: hyperspectral image processing allows for the use of advanced machine learning algorithms for classification, which can be trained to recognize complex patterns and features in the data. These algorithms can provide more accurate and robust classifications than traditional methods and can help identify even rare or subtle classes of materials or objects.
- High spatial resolution: hyperspectral imaging often has higher spatial resolution than multispectral technology, making it possible to accurately map and

monitor changes in an area of interest at a fine scale. This can be particularly important in applications such as mineral exploration, where precise mapping of mineral deposits is critical.

- Dimensionality reduction: hyperspectral data typically contains a large number of spectral bands, which can be computationally expensive to process. Hyperspectral image processing allows for the use of dimensionality reduction techniques, such as principal component analysis (PCA) or independent component analysis (ICA), which can reduce the number of spectral bands and improve computational efficiency.

- Feature extraction: hyperspectral image processing allows for the use of feature extraction techniques, such as the use of spectral indices, which can be used to extract relevant information from the data. This can be particularly useful in applications, such as agriculture, where specific spectral characteristics can indicate crops' health or stress levels.

- Time-series analysis: hyperspectral image processing allows for the analysis of time-series data, which can provide information on the dynamics of an area over time. This can be particularly useful in applications, such as coastal forest ecosystem monitoring, where the ability to track changes over time can provide valuable insights into ecosystem health and functioning.

- Cost-effective solution: hyperspectral image processing can be a cost-effective solution for analyzing large areas of data, especially when compared to traditional ground-based methods. The use of aircraft or satellite-based sensors can cover large areas in a short time and with minimal costs, providing valuable information for a wide range of applications.

Overall, hyperspectral image processing has the potential to significantly improve the accuracy and precision of various applications, such as mineral exploration, agriculture, forestry, military surveillance, and mangrove species classification. It can help overcome the limitations of traditional methods and provide a detailed understanding of the composition of the scene and the presence of rare or subtle classes of materials or objects, which can have a significant impact on different fields.

The potential of hyperspectral technology has already been successfully established in the field of vegetation research (Asner, 2000; Schmidt and Skidmore, 2003). Hyperspectral data has also opened novel opportunities for mapping mangrove forests, as with the improved level of detail, a description of the entire spectra of mangrove cover types is achievable (Green et al., 2000; Hirano et al., 2003). Measurements outside the visible wavelength range make possible new potential to distinguish between mangroves based on components, such as leaf water content, leaf chemistry with regard to the ecosystem, and ecological changes (Vaiphasa et al., 2005; Green et al., 1998). The capability to identify physiological strain circumstances by spectral response values for mangrove management (Long and Skewers, 1996) is of immense significance.

Image classification algorithms, minimum distance to mean (MDM), Mahalanobis distance (MD), maximum likelihood classifier (MLC), and spectral angle mapper (SAM) were implemented on hyperspectral data acquired by the sensor, airborne imaging spectrometer for applications (AISA+), to identify mangroves in two study areas of Texas, USA (Yang et al., 2009). In terms of overall classification accuracy, MD and MLC performed considerably well in relation to MDM and SAM (Yang et al., 2009).

Jia et al. (2014) used object-oriented methods on hyperspectral data of the Hyperion sensor and high spatial resolution data of the SPOT-5 sensor for species-level classification of mangroves in Mai Po Nature Reserve, Hong Kong.

Zeng et al. (2007) used linear spectral unmixing of high spatial resolution and hyperspectral data for geometric-optical modeling to retrieve a forest canopy variable (crown closure) in the broad-leaved natural forest reserve of the Three Gorges region of China.

Demuro and Chisholm (2003) applied a SAM classifier in the hyperspectral imagery of the Minnamurra River estuary in Australia, where they selected 105 noise-free bands. The classified image comprises nine non-vegetation classes, two mangrove species classes, and five additional vegetation classes. Mixture tuned matched filtering (MTMF) technique was applied by Demuro and Chisholm (2003) the results of which showed a similar distribution of mangrove species as observed in the field. Kamal and Phinn (2011) used hyperspectral data for a comparative analysis of mangrove species discrimination using pixel-based and object-based approaches.

Wang and Sousa (2009) developed a band-selection method for enhanced spectral discrimination of mangrove species at the leaf level. Their analysis was based on leaves of red, black, and white mangroves of Panama (USA), whose spectra were measured in the laboratory. They examined the discrimination ability by extracting six narrow bands at 780nm, 790nm, 800 nm, 1480 nm, 1530 nm, and 1550 nm that provided excellent spectral delineation.

Vaiphasa et al. (2005) carried out similar research with 16 mangrove species collected in Ao Sawi, Thailand. Here they have identified four spectral bands that distinguished these 16 mangrove species very efficiently, excluding species of the *Rhizophoraceae* family. In another study followed later, Vaiphasa (2005) applied a genetic search algorithm to categorize hyperspectral bands with maximum spectral discrimination. Each chosen band was directly connected to salient physio-chemical properties of vegetation, such as leaf pigments, internal leaf structure, and water content.

2.3 AUTOMATED ENDMEMBER DETECTION

Automated endmember detection is an essential step in the hyperspectral image processing workflow, as it allows for the identification of the pure spectral signatures or endmembers of the materials or objects present in the scene. This is essential for accurate spectral unmixing and classification of the data.

Several different algorithms and techniques can be used for automated endmember detection, including the following:

- Vertex component analysis (VCA): VCA is a popular unsupervised algorithm that uses a simplex-based approach to identify the endmembers in the data. It is computationally efficient and can handle high-dimensional data.

- N-FINDR: N-FINDR is another unsupervised algorithm that uses an iterative approach to identify the endmembers in the data. It is particularly useful for identifying endmembers in highly mixed pixels and can handle high-dimensional data.
- Minimum volume simplex analysis (MVSA): MVSA is an unsupervised algorithm that uses a volume-based approach to identify the endmembers in the data. It is particularly useful for identifying endmembers in highly mixed pixels and can handle high-dimensional data.
- Pixel purity index (PPI): PPI is a supervised algorithm that uses a pixel-based approach to identify the endmembers in the data. It is particularly useful for identifying endmembers in highly mixed pixels and can handle high-dimensional data.
- Deep learning-based endmember detection: another recent trend in endmember detection is using deep learning-based models like convolutional neural networks.

There has been significant research conducted worldwide on the topic of automated endmember detection in hyperspectral image processing.

Researchers have focused on developing and evaluating different algorithms and techniques for automated endmember detection. For example, researchers at the National Aeronautics and Space Administration (NASA) have developed and evaluated the performance of several algorithms, including VCA and N-FINDR. They have applied them to a variety of hyperspectral data sets. Researchers at the United States Geological Survey (USGS) have also developed and evaluated several algorithms, including MVSA and PPI. They have applied them to a variety of hyperspectral data sets.

Internationally, researchers have also focused on developing and evaluating different algorithms and techniques for automated endmember detection. For example, researchers in Europe have developed and evaluated the performance of several algorithms, including VCA and N-FINDR. They have applied them to a variety of hyperspectral data sets. Researchers in Asia have also developed and evaluated several algorithms, including MVSA and PPI. They have applied them to a variety of hyperspectral data sets.

In recent years, researchers have also applied deep learning techniques for endmember detection, these techniques have shown to be efficient and accurate in some applications.

Overall, there has been a significant amount of research conducted on the topic of automated endmember detection in hyperspectral image processing, both nationally and internationally. Many different algorithms and techniques have been developed and evaluated, and they have been applied to a wide range of hyperspectral data sets. This research has led to a better understanding of the strengths and weaknesses of different algorithms and techniques and has helped to improve the accuracy and precision of automated endmember detection in hyperspectral image processing.

In addition to the development and evaluation of different algorithms and techniques, researchers have also focused on the application of automated endmember detection to various fields such as mineral exploration, agriculture, forestry, military surveillance, and mangrove species classification. This research has shown the potential of automated endmember detection to improve the accuracy and precision of these applications and has led to the development of new methods and techniques for analyzing and interpreting the results.

Overall, the research on automated endmember detection in hyperspectral image processing has been very active, both nationally and internationally, and it continues to be an important area of research with many potential applications. The use of deep learning techniques has shown to be efficient and accurate in some applications, and it is expected that these techniques will continue to be explored in future research. These endmembers are used in the spectral unmixing process to estimate the abundance of each endmember in mixed pixels. There are several more algorithms and techniques that can be used for automated endmember detection, including the below:

- Linear mixing model (LMM) based algorithms: LMM-based algorithms, such as N-FINDR and VCA, use linear algebra to identify the endmembers in a data set. These algorithms are computationally efficient and work well for data sets with a limited number of endmembers. They are simple to implement and are based on a linear mixing model of the data, which assumes that the pixels in the scene are linear combinations of the endmembers.
- Non-linear mixing model (NLMM)-based algorithms: NLMM-based algorithms, such as MVSA and minimum volume convex hull (MVCH), use non-linear optimization techniques to identify the endmembers in a data set. These algorithms are more robust than LMM-based algorithms and can handle non-linear mixing, but they can be computationally expensive for large data sets.
- Sparsity-based algorithms: sparsity-based algorithms, such as orthogonal matching pursuit (OMP) and compressive sensing (CS), use mathematical models to identify the sparse representation of the data and can be used to identify endmembers in noisy or highly mixed data sets. These algorithms are based on the assumption that the data is sparse in some domains, and they exploit this sparsity to identify the endmembers.
- Clustering-based algorithms: clustering-based algorithms, such as K-means and expectation-maximization (EM), group similar pixels together and can be used to identify endmembers in data sets with a large number of endmembers. These algorithms are based on the assumption that the data is composed of clusters of pixels, where each cluster represents an endmember. They use statistical methods to group similar pixels together and identify the endmembers.
- Hybrid algorithms: hybrid algorithms, such as iteratively reweighted multivariate alteration detector (IR-MAD) and hybrid vertex component analysis (HVCA), combine different techniques and algorithms to improve the robustness and efficiency of endmember detection. These algorithms use a combination of LMM and NLMM techniques or a combination of different clustering and sparsity-based techniques to improve the results.

Each algorithm has its advantages and disadvantages, and the choice of algorithm will depend on the specific application, the characteristics of the data set, and the analysis goals. For example, LMM-based algorithms are simple to implement and are suitable for data sets with a limited number of endmembers, while NLMM-based algorithms can handle non-linear mixing, but they can be computationally expensive. Sparsity-based algorithms are suitable for noisy or highly mixed data sets, while clustering-based algorithms are suitable for data sets with a large number of endmembers. Hybrid algorithms can provide a balance between robustness and efficiency.

In summary, automated endmember detection is a crucial step in hyperspectral image processing, as it provides valuable information for spectral unmixing and other analysis tasks. Several algorithms and techniques are available for automated endmember detection, each with its advantages and disadvantages. The choice of algorithm will depend on the specific application, the characteristics of the data set, and the analysis goals.

2.4   SPECTRAL MIXTURE ANALYSIS

Spectral mixture analysis (SMA) is a powerful technique used to analyze and interpret the spectral information of hyperspectral images. It is a form of linear spectral unmixing that assumes that the pixels in the scene are linear combinations of pure spectral endmembers. Spectral mixture analysis can be used to estimate the abundance of different materials or objects in the scene, as well as their spectral signatures.

There are several different approaches to SMA, including the following:

- Linear spectral unmixing (LSU): this is the most basic form of SMA, which assumes that the pixels in the scene are linear combinations of pure spectral endmembers. LSU estimates the fractional abundance of each endmember in each pixel using a least-squares method.
- Non-negative matrix factorization (NMF): this approach is similar to LSU, but it also imposes non-negativity constraints on the abundance fractions, which can help prevent negative or complex solutions.
- Iterative endmember selection (IES): this approach uses an iterative process to identify and select the endmembers from the data. It starts with an initial set of endmembers and then iteratively refines the solution by updating the endmembers and abundances.
- Full spectral unmixing (FSU): this approach estimates the endmembers and abundances simultaneously, using a non-linear optimization method. It can handle cases where the number of endmembers is unknown or where the endmembers are not strictly non-negative.
- Convex/non-convex optimization: this approach uses a mathematical optimization method to estimate the endmembers and abundances in a data set, this can be convex optimization or non-convex optimization.

Spectral mixture analysis can be a powerful tool for extracting and analyzing spectral information from hyperspectral images, but it also has limitations. The assumptions of linear mixing and the presence of pure spectral endmembers may not always be accurate in real-world scenarios, and the choice of algorithm and endmember selection method can also affect the accuracy of the results. Additionally, SMA can be sensitive to noise and atmospheric effects, which can introduce errors in the estimated abundances.

In addition to these limitations, SMA is also sensitive to the number and quality of endmembers used in the analysis. The selection of endmembers can be a challenging task, and it can have a significant impact on the accuracy of the estimated abundances.

Despite these limitations, SMA remains a powerful technique for analyzing and interpreting the spectral information of hyperspectral images. It can provide valuable information on the composition and abundance of different materials or objects in the scene, and it can be used in a wide range of applications, such as mineral exploration, agriculture, and coastal ecosystem monitoring.

In conclusion, spectral mixture analysis is a useful technique for analyzing and interpreting the spectral information of hyperspectral images. It allows for the estimation of the abundance of different materials or objects in the scene and their spectral signatures, but it also has limitations. The assumptions of linear mixing and the presence of pure spectral endmembers may not always be accurate in real-world scenarios, and the choice of algorithm and endmember selection method can also affect the accuracy of the results. Despite these limitations, SMA remains a powerful technique for extracting and analyzing spectral information from hyperspectral images.

2.4.1 Linear Spectral Unmixing on Hyperspectral Data

LSU is a technique used to analyze and interpret the spectral information of hyperspectral images. It is a form of spectral mixture analysis that assumes that the pixels in the scene are linear combinations of pure spectral endmembers. LSU can be used to estimate the abundance of different materials or objects in the scene, as well as their spectral signatures.

LSU is based on LMM, which assumes that each pixel in the scene is a linear combination of the endmembers. The LMM can be represented mathematically as

$$y = Ax$$

where y is the vector of pixel spectra, A is the matrix of endmember spectra, and x is the vector of fractional abundances of the endmembers. LSU estimates the fractional abundances of the endmembers in each pixel using a least-squares method.

LSU has several advantages, including the following:

- simplicity and computational efficiency,
- robustness to noise and atmospheric effects,
- ability to handle a limited number of endmembers,
- it can be used for both supervised and unsupervised classification.

However, LSU also has several limitations, including the below:

- Assumes that the pixels in the scene are linear combinations of pure spectral endmembers, which may not always be accurate in real-world scenarios.

- Sensitive to the quality and number of endmembers used in the analysis.
- Assumes that the abundance fractions are non-negative, which may not always be the case.
- It can be sensitive to the presence of mixed pixels and outliers.

Despite these limitations, LSU remains a powerful technique for analyzing and interpreting the spectral information of hyperspectral images. It can provide valuable information on the composition and abundance of different materials or objects in the scene, and it can be used in a wide range of applications, such as mineral exploration, agriculture, and coastal ecosystem monitoring.

In addition to these limitations, LSU is also sensitive to the number and quality of endmembers used in the analysis. The selection of endmembers can be a challenging task, and it can have a significant impact on the accuracy of the estimated abundances. Therefore, it is important to carefully evaluate and select the endmembers for a given data set to achieve accurate results.

Recent developments in linear unmixing algorithms have been proposed to overcome the above limitations. For instance, sparsity-constrained linear unmixing, robust linear unmixing, and sparse linear unmixing.

Kanniah et al. (2007) applied spectral unmixing on IKONOS data and successfully identified three mangrove species in a mangrove forest in Malaysia. This model has also been applied for the abundance estimation of minerals (Kumar et al., 2001), identification of cropping patterns (Yang et al., 2007; Yang et al., 2010), classification of landcover (Jafari and Lewis, 2012; Shakoor, 2003), etc.

Graña and DAnjou (2005) applied principal component analysis to derive abundance estimates in spectral unmixing. Algorithms based on unconstrained regularization (Huck et al., 2010) have been developed for unmixing. The fully constrained least-squares (FCLS) algorithm (Heinz, 2001) is another method that represents the generalization of the least-squares formulation. Non-negative solutions with principles of greatest entropy have been proposed by Miao et al. (2007). There have been attempts to estimate abundances by means of a regularization framework in which various priors have been used on abundances to obtain improved results. A Bayesian outlook has been used by Bali and Djafari (2008), which imposes a volume prior and is based on the non-negative matrix factorization approach. Many algorithms use such types of priors that model spatial correlations that may exist between the abundance values (Eches et al., 2011), exclusive of their dependencies across the spectral space.

Dobigeon and Tourneret (2007) applied a hierarchical Bayesian model for spectral unmixing of hyperspectral data. The theory of endmember variability for the estimation of fractional abundances has been considered by Zare and Ho (2014). Unmixing based on total least-squares (TLS) approach has been proposed by Sirkeci et al. (2000), in which the solution has been restricted to yield physically constrained abundance values. Highly mixed pixels have been unmixed using neural networks (Licciardi and Del Frate, 2011). Abundances have also been estimated as linear sparse regression with the use of extracted endmembers in Bioucas-Dias et al. (2012).

LSU is a powerful technique for analyzing and interpreting the spectral information of hyperspectral images. It allows for the estimation of the abundance of different materials or objects in the scene and their spectral signatures. However, LSU also has limitations, including assumptions of linear mixing and the presence of pure spectral endmembers, which may not always be accurate in real-world scenarios, sensitivity to the quality and number of endmembers, and the assumption of non-negative abundance fractions. Despite these limitations, LSU remains a powerful technique for extracting and analyzing spectral information from hyperspectral images. With recent developments and advancements in linear unmixing algorithms, it is possible to overcome the limitations and achieve improved results.

2.4.2 Non-Linear Spectral Unmixing of Hyperspectral Data

Non-linear spectral unmixing (NLUS) is a technique used to analyze and interpret the spectral information of hyperspectral images. It is a form of spectral mixture analysis that aims to overcome the limitations of traditional linear spectral unmixing (LSU) methods. Unlike LSU, NLUS assumes that the pixels in the scene are non-linear combinations of pure spectral endmembers.

NLUS is based on NLMM, which assumes that each pixel in the scene is a non-linear combination of the end members. The NLMM can be represented mathematically as

$$y = f(A,x)$$

where y is the vector of pixel spectra, A is the matrix of endmember spectra, x is the vector of fractional abundances of the endmembers, and f is a non-linear function. NLUS estimates the fractional abundances of the endmembers in each pixel using non-linear optimization techniques.
NLUS has several advantages, including the following:

- ability to handle non-linear mixing,
- ability to handle a higher order of endmembers,
- ability to handle noise and atmospheric effects.

However, NLUS also has several limitations, including the below:

- More computationally expensive than LSU.
- Assumes that the pixels in the scene are non-linear combinations of pure spectral endmembers, which may not always be accurate in real-world scenarios.
- Sensitive to the quality and number of endmembers used in the analysis.
- Assumes that the abundance fractions are non-negative, which may not always be the case.
- It can be sensitive to the presence of mixed pixels and outliers.

Despite these limitations, NLUS is a powerful technique for analyzing and interpreting the spectral information of hyperspectral images. It can provide valuable information on the composition and abundance of different materials or objects in the scene, and it can be used in a wide range of applications, such as mineral exploration, agriculture, and coastal ecosystem monitoring.

This book has demonstrated the potential of hyperspectral data for reliable and detailed species-level classification of mangroves in remotely located forested islands of Sunderban. The Sunderbans are characterized by both 'pure' and 'mixed' forest patches (stands) that, respectively, include single/multiple morphological forms of mangrove species in a closed niche. Pure mangrove patches could be subjected to sub-pixel classification through LSU, where light is reflected linearly from objects with minimum multiple scattering (Keshava and Mustard, 2002; Kanniah et al., 2007; Kanniah, 2005). However, in natural forests like Sunderbans, multiple scattering between vegetation canopies becomes significantly non-linear (Nascimento and Bioucas-Dias, 2009, 2010; Fan et al., 2009). The random distribution of endmembers produces nonlinearities in photon interactions, which are better characterized using non-linear spectral unmixing models. These models take into account interactive radiations between different endmembers lying in close proximity in a natural system; this is neglected in the case of linear spectral unmixing models. Non-linear models are thus expected to provide a more accurate representation of fractional abundance estimates of endmembers and their interactive mixtures.

Non-linear models have been developed for entities that are intimately mixed (Hapke, 1993). Such admixtures are typical images of sand or mineral deposits. Based on ray tracing (RT) theory, several theoretical frameworks have been derived to accurately describe interactions of light on encountering surfaces composed of particles (Somers et al., 2014).

Another type of non-linear interaction occurring at the macroscopic scale has also been mentioned, particularly in multilayered configurations. This model is encountered when light scattered by a given entity reflects off from other materials before reaching the sensor. This is often the case in forested areas, where there are many interactions between the ground and the canopy. Nascimento's model and Fan's model have been developed to analytically describe these interactions (Nascimento and Bioucas-Dias, 2009, 2010; Fan et al., 2009). These, however, pertain to second-order interactions with orders greater than two neglected. However, in natural Sunderban forests with mixed mangrove patches, more than two species exist in mixtures of mangroves within a pixel area. Models exist which consider higher interactions between different endmembers up to the 'nth' to overcome the limitations of previous models.

Moreover, these existing models only include inter-component interactions with no provision of intra-component interactions. As the target endmembers may occur in pure patches, multiple scattering between similar endmembers are present and should not be ignored. The model includes the interaction of similar endmembers to get a more accurate representation of abundances and final reflectance values.

In conclusion, non-linear spectral unmixing (NLUS) is a powerful technique for analyzing and interpreting the spectral information of hyperspectral images. It allows for the estimation of the abundance of different materials or objects in the scene and their spectral signatures, and it can handle non-linear mixing. However, NLUS also has limitations, including assumptions of non-linear mixing, sensitivity to the quality and number of endmembers, and the assumption of non-negative abundance fractions. Despite these limitations, NLUS remains a powerful technique for extracting and analyzing spectral information from hyperspectral images.

3 Preprocessing of Data

3.1 PREPROCESSING OF DATA

Preprocessing hyperspectral data is a crucial step in analyzing and interpreting the data. It is preparing and cleaning the raw data obtained from the hyperspectral sensor before any further analysis is performed. Preprocessing of hyperspectral data is a multistep process that includes several steps, such as atmospheric correction, noise reduction, geometric correction, and ground survey.

1. Atmospheric correction: this step is used to correct the effects of the atmosphere on the data, such as absorption and scattering. This is typically done using atmospheric models, such as the atmospheric radiative transfer model (ARTM), which can estimate the atmospheric conditions at the time of data acquisition.
2. Noise reduction: this step removes noise from the data, such as electronic and speckle noise. This can be done using various techniques such as spatial filtering, principal component analysis (PCA), or independent component analysis (ICA).
3. Geometric correction: this step is used to correct for geometric distortions in the data, such as those caused by the sensor's position or the terrain. This can be done using techniques, such as image registration, which aligns multiple images taken at different times or from different angles, and geometric correction, which corrects for distortions caused by the sensor's position or the terrain.
4. Ground survey: this step collects information about the scene before acquiring the hyperspectral data. This information can be used to correct geometric distortions, atmospheric effects, and other sources of error in the hyperspectral data. The ground survey can be conducted using a variety of methods, such as ground-based spectroscopy, field sampling, and ground-based imaging.

The importance of preprocessing step in hyperspectral data analysis is that it ensures that the data is of high quality and accurate, it also allows for the removal of any artifacts or errors present in the data, which can improve the performance of the subsequent analysis. Furthermore, preprocessing also ensures that the data is in a format

DOI: 10.1201/9781003432623-3

that is compatible with the intended analysis techniques, making it easier to work with and interpret.

In the context of this book, which focuses on mangrove species classification, preprocessing is particularly important. Mangrove species can have distinct spectral characteristics, and it is crucial to ensure that the data is of high quality and accuracy to accurately identify and classify the different species. The ground survey is critical in this case as it provides valuable information about the scene, such as the location and distribution of different mangrove species, which can be used to improve the overall performance of the analysis.

The preprocessing steps are essential to ensure that the data is of high quality and accurate and that it is in a format that is compatible with the intended analysis techniques. In the context of this book, which focuses on mangrove species classification, preprocessing is particularly important to accurately identify and classify different species.

3.2 GROUND SURVEY

The ground survey is an essential step in the preprocessing of hyperspectral data, it is typically carried out in the following steps:

1. Planning: the first step in the ground survey is to plan the survey by identifying the area of interest, the objectives of the survey, and the equipment and personnel required. This includes determining the location of the survey, the time of year, and the weather conditions.
2. Equipment preparation: once the plan is in place, the equipment is prepared for the survey. This includes checking and calibrating the instruments and cameras and ensuring that they are in good working condition.
3. Data collection: the data collection phase is the most critical step in the ground survey. This includes capturing the spectral and spatial information of the scene using ground-based spectroscopy, field sampling, and ground-based imaging. The data should be collected under optimal conditions, such as a clear sky and low sun elevation.
4. Data processing: the data is processed after the data collection phase. This includes correcting for geometric distortions, atmospheric effects, and other sources of error in the data. The data processing step can be time-consuming and requires specialized software and expertise.
5. Data analysis: the processed data is then analyzed to extract useful information about the scene. This includes identifying and classifying different materials or objects in the scene using various techniques, such as principal component analysis (PCA) or independent component analysis (ICA).
6. Data interpretation: the final step is to interpret the results of the data analysis. This includes generating maps or other visualizations to display the results and performing further analysis to extract more detailed information about the materials or substances identified in the images.

7. Quality assurance: it is essential to ensure the quality and accuracy of the data. It includes checking the data for errors and inconsistencies and comparing it to other data sources, such as aerial photographs, maps, or previous surveys.

Overall, the ground survey is a multistep process that requires careful planning, specialized equipment, and expertise to ensure the accuracy and quality of the hyperspectral data. It is a crucial step in the preprocessing of hyperspectral data, providing valuable information about the scene that can be used to correct for errors and inaccuracies in the data and to identify and classify different materials or objects in the scene. The ground survey data is also useful for later stages of analysis and interpretation of hyperspectral data, such as for creating ground truth for supervised classification or for creating reference data for endmember extraction.

It is important to note that ground survey can be time-consuming and costly, and it is not always possible or necessary for all applications. However, for applications such as mangrove species classification, where the identification of different species is critical, and their spectral characteristics may vary, the ground survey is crucial for providing accurate information about the scene and improving the analysis's overall performance.

Ground survey for mangrove species classification typically involves a combination of field sampling, ground-based spectroscopy, and ground-based imaging. The specific steps will depend on the specific research goals and the characteristics of the mangrove ecosystem being studied.

1. Field sampling involves collecting samples of mangrove species in the field and analyzing them in a laboratory. This can provide information about the chemical and physical properties of the species, such as the presence of specific pigments or compounds, which can be used to identify and classify different species.
2. Ground-based spectroscopy: this involves collecting reflectance or emission spectra of the mangrove species using a portable spectrometer. This can provide information about the spectral characteristics of the species, such as the presence of specific absorption or reflection bands, which can be used to identify and classify different species.
3. Ground-based imaging involves capturing mangrove species' images using a digital camera or other imaging equipment. This can provide information about the spatial characteristics of the species, such as the size, shape, and distribution of different species, which can be used to improve the overall performance of the analysis.
4. Data collection: the data collection phase is the most critical step in the ground survey. This includes capturing the spectral and spatial information of the scene using ground-based spectroscopy, field sampling, and ground-based imaging. The data should be collected under optimal conditions, such as a clear sky and low sun elevation.
5. Data analysis: the processed data is then analyzed to extract useful information about the scene. This includes identifying and classifying different

mangrove species in the scene using various techniques, such as principal component analysis (PCA) or independent component analysis (ICA).

6. Data interpretation: the final step is to interpret the results of the data analysis. This includes generating maps or other visualizations to display the results and performing further analysis to extract more detailed information about the mangrove species identified in the images.

3.2.1 CASE STUDY-BASED GROUND TRUTH RESULTS

Ground survey of the study area was conducted to identify and collect samples of mangrove species whose image-based classification has been carried out. This was made in June 2011 on the immediate acquisition of data. GPS (Global Positioning System of make GARMIN GPSMAP 78s with high-performance marine handheld three-axis compass and barometric altimeter and accuracy of 4m) has been used to precisely locate the geographical coordinates of the study trail. The field quadrat survey-cum-sample method was adopted to assess the frequency of species dominance. The quadrat size was taken as 30m × 30m, which equals the spatial resolution of Hyperion imagery. Thirty quadrat plots were selected and examined in pure and mixed patch jungles that were henceforth plotted on geometrically corrected hyperspectral imagery with the help of GPS. The individuals of different species present within each quadrat were counted, and samples were taken. The species with more than 50% abundance was considered as 'dominant,' and the corresponding patch represents a pure stand. The quadrats in which none of the mangrove species attain 50% dominance are designated a mixed patch. These ground survey areas are used as reference points for the accuracy assessment of image-derived findings (Table 3.1). The ground truthing exercise was again repeated in May 2012 and June 2013. No significant change in the physical abundance of individual plants could be noted within this time span.

Overall, the ground survey is a multistep process that requires careful planning, specialized equipment, and expertise to ensure the accuracy and quality of the hyperspectral data. It is a crucial step in preprocessing hyperspectral data, providing valuable information about the scene that can be used to identify and classify different mangrove species.

3.3 SPECTRAL LIBRARY DEVELOPMENT FROM GROUND DATA

It creates a collection of reference spectra for different materials or objects in a scene using ground data collected from the scene. This process is typically used in hyperspectral image processing to identify and classify different materials or objects in the scene.

1. Ground data collection: the first step in spectral library development is to collect ground data from the scene. This typically involves collecting reflectance or emission spectra of different materials or objects in the scene using ground-based spectroscopy, field sampling, or ground-based imaging. The

TABLE 3.1

Results of Quadrat-Based Sample Survey Conducted in the Study Area Size of Each Quadrat: 30m × 30m area

Sl.No.	Geographical Co-ordinates	Dominant Mangrove Species Identified within Each Quadrat	Area (in %) covered by the Dominant Species within each Quadrat	Type of Mangrove Patch Identified (Pure/ Mixed)
1.	21°34.621'N; 88°16.502'E	*Excoecaria agallocha*	70	Pure
2.	21°34.626'N; 88°16.499'E	*Excoecaria agallocha*	70	Pure
3.	21°34.630'N; 88°16.490' E	*Excoecaria agallocha*	72	Pure
4.	21°34.616'N; 88°16.513'E	*Excoecaria agallocha*	65	Pure
5.	21°34.650' N; 88°16.788' E	*Avicennia marina*	80	Pure
6.	21°34.642' N; 88°16.799' E	*Avicennia marina*	82	Pure
7.	21°34.631' N; 88°16.810' E	*Avicennia marina*	78	Pure
8.	21°34.621' N; 88°16.823' E	*Avicennia marina*	90	Pure
9.	21o34.380' N; 88o17.807' E	*Ceriops decandra* *Excoecaria agallocha* *Phoenix paludosa* *Avicennia marina*	45 25 10 30	Mixed
10.	21°34.385' N; 88°17.811' E	*Ceriops decandra* *Excoecaria agallocha* *Phoenix paludosa* *Avicennia marina*	45 30 10 20	Mixed
11.	21°34.374' N; 88°16.809' E	*Ceriops tagal* *Excoecaria agallocha* *Phoenix paludosa* *Avicennia marina*	30 21 19 30	Mixed
12.	21°34.374' N; 88°16.812' E	*Ceriops tagal* *Excoecaria agallocha* *Phoenix paludosa* *Avicennia marina*	40 30 20 10	Mixed

data should be collected under optimal conditions, such as a clear sky and low sun elevation.

2. Data preparation: once the ground data is collected, it needs to be prepared for further analysis. This includes removing any artifacts or errors present in the data, such as noise or atmospheric effects, and converting the data into a format that is compatible with the intended analysis techniques.

3. Endmember extraction: the next step is to extract the endmembers from the ground data. Endmembers are the pure spectral signatures of the materials or objects in the scene. This can be done using various techniques, such as the N-FINDR algorithm or the vertex component analysis (VCA) algorithm.

4. Spectral library development: once the endmembers are extracted, they can be used to create a spectral library. This typically involves creating a collection of reference spectra for different materials or objects in the scene, along with information about the location and distribution of the materials or objects in the scene.

5. Validation: the final step is to validate the spectral library. This includes comparing the reference spectra in the library with the original ground data to ensure that the library is accurate and reliable. It also includes testing the library against other data sets to ensure that it can be used to identify and classify different materials or objects in other scenes.

Overall, the development of a spectral library from ground data is creating a collection of reference spectra for different materials or objects in a scene using ground data collected from the scene. This process is an essential step in hyperspectral image processing, providing valuable information about the scene that can be used to identify and classify different materials or objects in the scene. However, it is essential to note that spectral library development can be time-consuming and costly, and it requires specialized equipment and expertise.

3.3.1 Case Study-Based Results

From the ground survey, it was possible to achieve precise identification and collection of ground control points of dominant mangrove species (Figure 3.1). The spectral signature of each of the dominant species hitherto identified was plotted on the Hyperion image (Table 3.2). Figure 3.2 displays the spectral library of these dominant mangrove species, namely *Excoecaria agallocha*, *Avicennia officinalis*, *Ceriops decandra*, *Avicennia marina*, *Phoenix paludosa*, *Bruguiera cylindrica*, and *Avicennia alba*. Table 3.3 displays the location details of ground control points of pure mangrove patches in the study area.

The reflectance of *Excoecaria agallocha* is lower compared to the other six species throughout the visible, near infra-red (NIR), and short wave infra-red (SWIR) regions. This is due to the high chlorophyll and water content of its leaves. The highest reflectance is that of *Phoenix paludosa*, *Avicennia marina*, and *Avicennia officinalis* owing to their leaf color and canopy density. The species are quite differentiable in the NIR region, where reflectance is essentially due to the internal structure of the leaf and the radiation that penetrates it (Cochrane, 2000; Curran, 1989). In the SWIR region, the

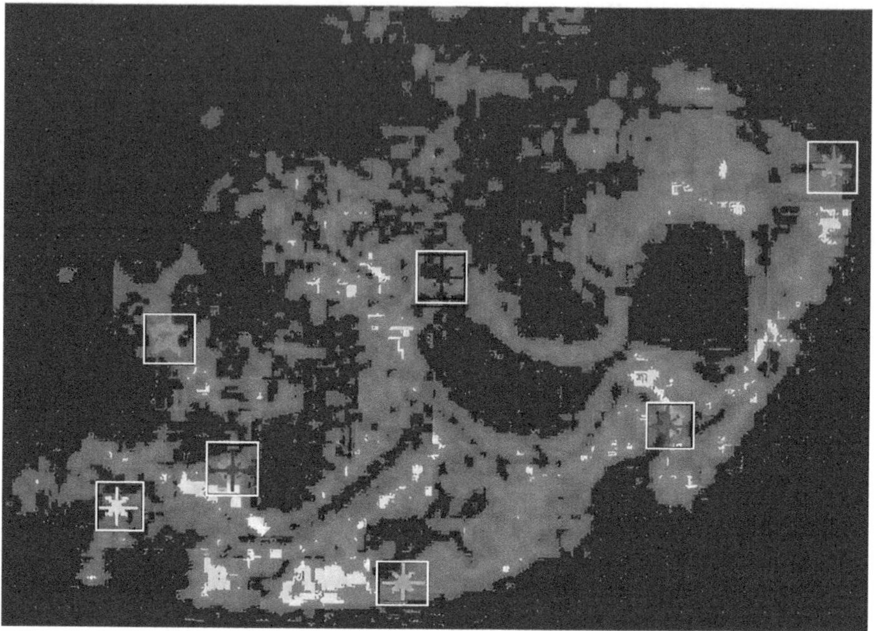

FIGURE 3.1 Plotted locations of mangrove species collected during ground survey.

spectral curves are quite distinct. This is the hydric zone where the amount of leaf water affects the pattern of spectral reflectance. It is observed that the reflectance of mangrove species in this band is lower than non-mangrove vegetation. The low reflectance of mangrove leaves is attributed to weaker scattering from intercellular air spaces compared to non-mangrove leaves (Elvidge, 1987). Low reflectance of mud and water in SWIR bands may further reduce the reflected radiance of mangrove forests as a whole.

3.4 PREPROCESSING OF HYPERSPECTRAL DATA

3.4.1 ACQUISITION OF HYPERSPECTRAL DATA

Acquiring hyperspectral data is a critical step in analyzing and interpreting hyperspectral images. It involves capturing the reflectance of a scene at many different wavelengths of light, resulting in a three-dimensional data cube with the spatial information in the x and y dimensions and the spectral information in the z dimension.

1. Data acquisition: hyperspectral images are typically acquired using specialized sensors mounted on aircraft or satellites. These sensors measure the reflectance of different wavelengths of light across a wide range of the electromagnetic spectrum and generate a series of images, one for each wavelength band. The acquisition is made based on the following methods:

TABLE 3.2

Results of Quadrat-Based Sample Survey Conducted in the Study Area Size of Each Quadrat: 30m × 30m area

Sl. No.	Geographical Co-ordinates	Dominant Mangrove Species Identified within Each Quadrat	Area (in %) covered by the Dominant Species within each Quadrat	Type of Mangrove Patch Identified (Pure/Mixed)
1.	21°34.621'N; 88°16.502'E	*Excoecaria agallocha*	70	Pure
2.	21°34.626'N; 88°16.499'E	*Excoecaria agallocha*	70	Pure
3.	21°34.630'N; 88°16.490' E	*Excoecaria agallocha*	72	Pure
4.	21°34.616'N; 88°16.513'E	*Excoecaria agallocha*	65	Pure
5.	21°34.650' N; 88°16.788' E	*Avicennia marina*	80	Pure
6.	21°34.642' N; 88°16.799' E	*Avicennia marina*	82	Pure
7.	21°34.631' N; 88°16.810' E	*Avicennia marina*	78	Pure
8.	21°34.621' N; 88°16.823' E	*Avicennia marina*	90	Pure
9.	21°34.380' N; 88°17.807' E	*Ceriops decandra* *Excoecaria agallocha* *Phoenix paludosa* *Avicennia marina*	45 25 10 30	Mixed
10.	21°34.385' N; 88°17.811' E	*Ceriops decandra* *Excoecaria agallocha* *Phoenix paludosa* *Avicennia marina*	45 30 10 20	Mixed
11.	21°34.374' N; 88°16.809' E	*Ceriops tagal* *Excoecaria agallocha* *Phoenix paludosa* *Avicennia marina*	30 21 19 30	Mixed
12.	21°34.374' N; 88°16.812' E	*Ceriops tagal* *Excoecaria agallocha* *Phoenix paludosa* *Avicennia marina*	40 30 20 10	Mixed

TABLE 3.3
Location Details of Ground Control Points of Pure Mangrove Patches in the Study Area

Name of Mangrove Species Identified	Latitude (N)	Longitude (E)
Excoecaria agallocha	21°34'9.40"	88°17'10.68"
Ceriops decandra	21°34'48.99"	88°18'3.18"
Phoenix paludosa	21°34'3.17"	88°15'52.40"
Avicennia alba	21°33'39.46"	88°16'30.78"
Avicennia marina	21°34'13.15"	88°17'30.53"
Bruguiera cylindrica	21°33'55.56"	88°15'26.26"
Avicennia officinalis	21°34'31.57"	88°15'38.04"

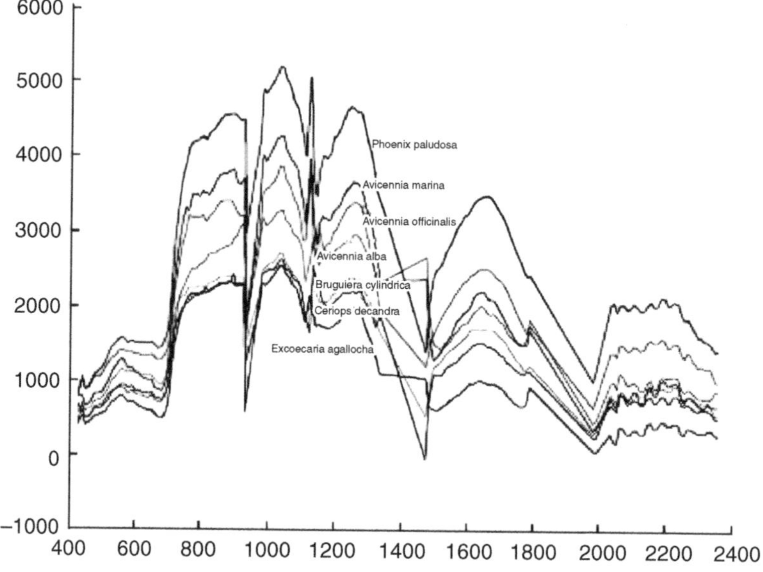

FIGURE 3.2   Spectral library of seven mangrove species dominant in Henry Island.

a. Satellite-based data acquisition: this is one of the most common ways of acquiring hyperspectral data. Several commercial and government satellites collect hyperspectral data, such as the Hyperion sensor on NASA's EO-1 satellite, the AVIRIS sensor on NASA's AVIRIS aircraft, and the Hyperion sensor on the Indian ResourceSat-2A satellite. These sensors collect data across a wide range of the electromagnetic spectrum, including the visible and near-infrared ranges. They are capable of collecting data with high spatial and spectral resolution.

b. Airborne-based data acquisition: another way to acquire hyperspectral data is through the use of airborne sensors. This can be done using aircraft or UAVs (unmanned aerial vehicles). Examples include the AVIRIS sensor on NASA's AVIRIS aircraft and the HyMap sensor on the CSIRO's AASI aircraft. These sensors can collect data with high spatial resolution and can collect data over a wide range of the electromagnetic spectrum.

c. Handheld data acquisition: handheld spectrometers can be used to acquire hyperspectral data in the field. These devices are portable and can be used to collect data in difficult-to-access areas, such as remote forests or wetlands. Examples include the ASD FieldSpec HandHeld 2 and the Ocean Optics USB2000+. These sensors collect data in a limited number of bands and can be helpful in the field of identifying the species.

d. Ground-based data acquisition: ground-based spectrometers can be used to collect reflectance spectra of the samples in the field. These devices can be used to create a ground spectral library, which can be used for comparison with the hyperspectral data and training and validation of the classification algorithms. Examples include the ASD FieldSpec Pro and the Analytical Spectral Devices FieldSpec 4.

2. Data collection strategies: the strategy will depend on the specific application and area of interest. Government agencies such as NASA and the US Geological Survey (USGS) collect and make hyperspectral data open to the public. These data are often collected using satellite or aircraft-based sensors and cover various locations and applications. Commercial providers offer hyperspectral data for a fee. These data may be collected using satellite or aircraft-based sensors or through partnerships with government agencies. Research institutions and universities also collect and make hyperspectral data available to the public.

3. Data quality: data quality is an important aspect to consider when acquiring hyperspectral data. Factors such as spatial resolution, spectral resolution, signal-to-noise ratio, and radiometric accuracy can affect the overall quality of the data. It is essential to carefully evaluate the quality of the data before using it for analysis and interpretation.

Case Study-Based Hyperspectral Image Acquisition

The present chapter makes use of hyperspectral data acquired by the Hyperion sensor on board the Earth Observatory-1 (EO-1) satellite. To acquire this data, a data acquisition request (DAR) was formally made to the United States Geological Survey (USGS)-Earth Observatory (EROS) Data Center requesting the image of Henry Island of the Sunderban forest. Accordingly, cloud-free data was acquired on 27 May 2011 and stored in their database. The Level 1R (L1R) Hyperion data in HDF (hierarchical data format) was downloaded in which the data obtained are in the form of the image file (with .L1R extension), an image header file, and a .met file. The .met file contains metadata with information such as image acquisition date, time, latitude–longitude, etc. Figure 3.3 shows the browse overlay of the acquired Hyperion image over the Sunderban region. The metadata for the acquired imagery is given in Table 3.4.

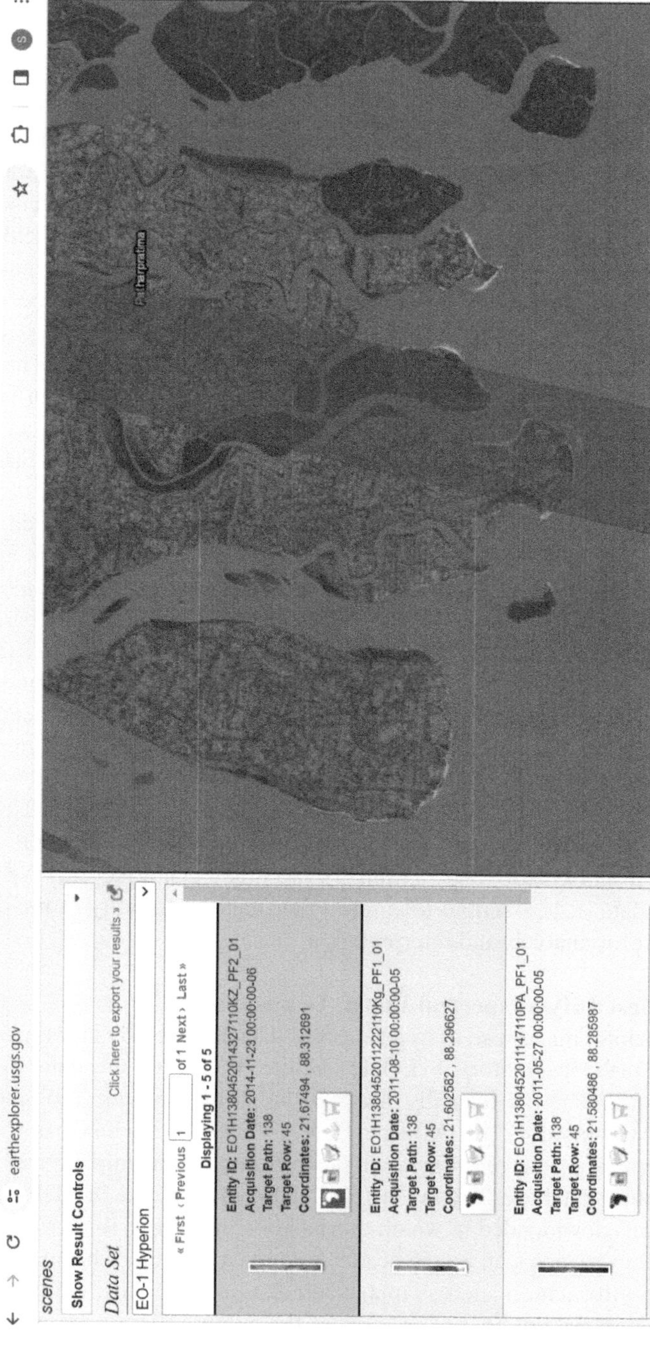

FIGURE 3.3 Browse overlay of Hyperion image of the Sunderban rainforest.

TABLE 3.4
Metadata of the Acquired Imagery

Data Set Attribute	Attribute Value
Entity ID	EO1H1380452011147110PA_PF1_01
Acquisition Date	2011/05/27
NW Corner	22°00'41.99"N; 88°20'55.62"E
NE Corner	21°59'50.47"N; 88°25'10.02"E
SW Corner	21°09'48.92"N; 88°09'09.83"E
SE Corner	21°08'57.60"N; 88°13'22.74"E
Image Cloud Cover	30% to 39% Cloud Cover
Receiving Station	PF1
Scene Start Time	2011:147:04:18:06.243
Scene Stop Time	2011:147:04:18:21.243
Date Entered	2011/05/27
Target Path	138
Target Row	45
Orbit Path	138
Sun Azimuth	85.895254
Sun Elevation	65.396135
Satellite Inclination	98.08
Orbit Row	45
Look Angle	−2.1333
Browse Available	Y

4. Ground survey: in addition to data acquisition from satellites or aircraft, it is also essential to conduct a ground survey to validate the results of the hyperspectral data analysis and to develop a ground spectral library. During the ground survey, field measurements are made using a field spectroradiometer or other spectral measurement device to collect reflectance spectra of the various mangrove species. These measurements are used to create a ground spectral library, which can be used for comparison with the hyperspectral data and training and validation of the classification algorithms.

5. Sampling strategy: a well-designed sampling strategy is crucial to ensure that the data collected is representative of the area of interest and that the results of the analysis are reliable. In the case of mangrove species classification, the sampling strategy should ensure that a diverse range of mangrove species and environmental conditions are represented in the data. This may include sampling at different times of the year, at different locations, and different scales.

In this book, the focus is on the classification of mangrove species, and thus the data acquisition process should be tailored to this specific task. The data should be collected in a way that allows for the identification and classification of different mangrove species within the scene. This may include collecting data under optimal conditions, such as a clear sky and low sun elevation, and using sensors with high spatial and spectral resolution.

Overall, the acquisition of hyperspectral data is a critical step in the analysis and interpretation of hyperspectral images. The data should be collected using specialized sensors, with appropriate data collection strategies and ground survey, with a well-designed sampling strategy to ensure that the data is representative of the area of interest and that the results of the analysis are reliable.

3.4.2 Bands and Spectral Ranges for Hyperspectral Data

Bands and spectral ranges are essential considerations when working with hyperspectral data. The spectral range refers to the range of wavelengths of light captured by the sensor, while the bands refer to the specific wavelength ranges within the spectral range that are measured by the sensor.

1. Spectral range: hyperspectral sensors typically measure the reflectance of a scene across a wide range of the electromagnetic spectrum, typically in the visible and near-infrared ranges. The specific spectral range depends on the sensor and can vary from sensor to sensor. For example, the AVIRIS sensor on NASA's AVIRIS aircraft measures the reflectance of a scene across the spectral range of 400 nm to 2500 nm, while the Hyperion sensor on NASA's EO-1 satellite measures the reflectance of a scene across the spectral range of 400 nm to 2500 nm.

2. Bands: the bands refer to the specific wavelength ranges within the spectral range that are measured by the sensor. Hyperspectral sensors typically measure the reflectance of a scene across hundreds or thousands of narrow, contiguous bands. The specific number of bands and their widths depend on the sensor and can vary from sensor to sensor. For example, the AVIRIS sensor on NASA's AVIRIS aircraft measures 224 bands with a width of 10 nm, while the Hyperion sensor on NASA's EO-1 satellite measures 256 bands with a width of 10 nm.

3. Spectral resolution: spectral resolution is the ability of a sensor to distinguish between different wavelengths of light. It is usually measured in nanometers (nm) and depends on the width of the bands. A sensor with a higher spectral resolution will have narrower bands, which allows it to distinguish between closely spaced wavelengths of light. A sensor with a lower spectral resolution will have wider bands, which will not allow it to distinguish between closely spaced wavelengths of light.

The choice of spectral range and bands can significantly impact the analysis and interpretation of hyperspectral data. For example, a sensor with a wide spectral range will allow for the detection of a more significant number of materials and substances. In contrast, a sensor with a narrow spectral range will only allow for the detection of a limited number of materials and substances. Similarly, a sensor with a large number of bands will allow for a greater detail in the data, while a sensor with a small number of bands will only allow for a limited detail in the data.

Case Study-Based Result

The Level-1 Hyperion data consists of 242 nominal bands covering the wavelengths from 356 nm to 2577 nm. The very near infra-red (VNIR) detector collects data in bands 1 to 70 and the short wave infra-red (SWIR) detector collects data from bands 71 to 242. There is an overlap between the VNIR and SWIR regions. Some of the 242 spectral channels are not calibrated in Level-1 data. There are channels that were not activated in the Hyperion arrays, some have low signal levels, and some are duplicated in VNIR-SWIR overlap region. The Level-1 data contains 200 calibrated channels of which 196 are unique. The digital values of the Level-1 product are 16-bit radiances that are stored as a 16-bit signed integer. The Level 1 product is only radiometrically corrected and is not atmospherically and geometrically resampled.

Overall, bands and spectral ranges are essential considerations when working with hyperspectral data. The choice of spectral range and bands can significantly impact the analysis and interpretation of the data, and it is vital to choose a sensor that is appropriate for the specific application and the area of interest.

3.4.3 REMOVAL OF ABSORPTION BANDS AND BANDS HAVING NO INFORMATION

Removing absorption bands and bands having no information is an important step in preprocessing hyperspectral data, as it helps to improve the accuracy and robustness of the results. This can be done using a variety of methods, such as atmospheric correction, noise reduction, and band selection.

1. Atmospheric correction: the atmospheric conditions can affect the accuracy of the hyperspectral data by introducing absorption bands in the data. These bands can be caused by gases such as water vapor, carbon dioxide, and ozone, as well as by aerosols, such as dust and smoke. To correct for these effects, atmospheric correction algorithms can be used. One of the widely used algorithms is the FLAASH (fast line-of-sight atmospheric analysis of spectral hypercubes) algorithm, which combines radiative transfer models, atmospheric profiles, and ground reflectance measurements to correct atmospheric effects and remove absorption bands.

2. Noise reduction: noise can be introduced into the hyperspectral data during the data acquisition process and can affect the accuracy of the data. Various noise reduction techniques can be used to reduce the noise, such as smoothing, median filtering, and principal component analysis (PCA). For example, applying a median filter to the data can effectively reduce noise caused by detector noise, electronic noise, and other random noise sources.

3. Band selection: selecting the appropriate bands for the analysis is important for the accuracy of the results. Bands can be selected based on their relevance to the application and the area of interest. For example, in a study of mangrove species classification, bands in the near-infrared region of the spectrum are more relevant as they contain information about the chlorophyll content of the plants. Bands that are not relevant or contain no data can be removed. This

can be done using various band selection methods, such as principal component analysis (PCA). It has been found that some bands are set to zero during Level-1 processing.

Case Study-Based Results

The zeroed bands are 1 to 7, 58 to 76 and 225 to 242. Bands 56 and 57 in the VNIR overlap with bands 77 and 78 in the SWIR region. Bands 77 and 78 have usually been found to be noisier than the corresponding bands in the VNIR. As the zero data bands and the VNIR-SWIR overlap (bands 56, 57 or 77, 78) are ignored, there are 196 unique calibrated bands (bands 8 to 57 and bands 79 to 224), which can be retained for processing. Within the 196 unique bands there exist a number of atmospheric water vapor bands that absorb most of the solar radiation. These have been identified by examining the radiance spectra. While analyzing the hyperspectral images, it has been found that the strongest water vapor bands occur between 1346nm (band 120) to 1497nm (band135), 1517nm (band 137) to 1537nm (band 139), 1568nm (band 142) to 1578nm (band 143), 1598nm (band 145), 1669nm (band 152) to 1689nm (band 154), 1709nm (band 156) to 2385nm (band 213). Since these bands contain little or no information, they have been ignored during further processing (Beck, 2003).

However, ENVI-FLAASH requires the bands centered near 1380nm (that represent the strong water vapor wavelengths) for masking of high altitude clouds. Therefore, bands 123 to 125 (covering 1376nm, 1386nm, and 1396 nm) have been retained on the image. Hence 156 hyperspectral bands have been ultimately used for further processing. This reduces data volume and speed-up the procedure. It is essential to carefully evaluate the hyperspectral data and remove absorption bands and bands having no information, as it can significantly improve the accuracy and robustness of the results. For example, if atmospheric correction is not applied, the absorption bands caused by atmospheric gases can lead to the misclassification of materials or substances in the scene. Similarly, if noise reduction is not applied, the noise can lead to inaccurate results. Band selection can also improve the accuracy of the results as it helps to remove bands that are not relevant or contain no information. These methods can be applied separately or in combination, depending on the specific application and the analysis goals.

3.4.4 REMOVAL OF BAD COLUMNS AND VERTICAL STRIPES

Removing bad columns and vertical stripes from hyperspectral data is an important step in preprocessing as it helps to improve the quality and accuracy of the data.

1. Removal of bad columns: bad columns are columns in the data that contain errors or anomalies, such as noise or missing data. These columns can be caused by various factors, such as detector malfunctions or electronic errors. One example of removing bad columns is using statistical analysis. A simple method is to calculate the standard deviation of each column, and then remove columns that have a standard deviation above a certain threshold. Another

example is using interpolation methods, such as linear or polynomial interpolation to estimate the value of missing or corrupted pixels.

2. Removal of vertical stripes: vertical stripes are horizontal lines in the data that contain errors or anomalies, such as noise or missing data. These stripes can be caused by various factors, such as detector malfunctions or electronic errors. To remove vertical stripes, various methods can be used, such as statistical analysis, interpolation, and outlier detection. For example, a simple method is to remove lines that have a standard deviation above a certain threshold.

Removing bad columns and vertical stripes from the data is important as it can greatly improve the quality and accuracy of the data. These errors can cause misclassification of materials or substances in the scene and can lead to inaccurate results. It is essential to remove these errors before further analysis and interpretation of the data.

3.5 ATMOSPHERIC CORRECTION

This is an important step in preprocessing of hyperspectral data as it helps to remove the effects of the atmosphere on the signal, which can introduce errors and noise into the data. There are several methods available for atmospheric correction, each with their own advantages and limitations.

1. Radiative transfer models: radiative transfer models are one of the most common methods for atmospheric correction. These models use information about the atmosphere, such as atmospheric profiles and ground reflectance measurements, to simulate the effects of the atmosphere on the signal. The most widely used radiative transfer models are the 6S model and the FLAASH (fast line-of-sight atmospheric analysis of spectral hypercubes) algorithm. The 6S model is a widely used model that can be used to correct for atmospheric effects by simulating the radiance and reflectance at the sensor. The FLAASH algorithm uses a combination of radiative transfer models, atmospheric profiles, and ground reflectance measurements to correct for atmospheric effects and remove absorption bands.

2. Dark object subtraction: another method of atmospheric correction is the dark object subtraction method, also known as the DOS method. This method is based on the principle that a dark object, such as a shadowed area, should have a reflectance of zero in the absence of atmospheric effects. This method uses the reflectance of a dark object to estimate the atmospheric effects and remove them from the data.

3. QUAC (quick atmospheric correction): QUAC is an atmospheric correction algorithm that uses a fast and efficient method for estimating the atmospheric effects on the signal. QUAC uses a simple, physically based model of the atmosphere and a set of look-up tables to estimate the atmospheric effects on the signal.

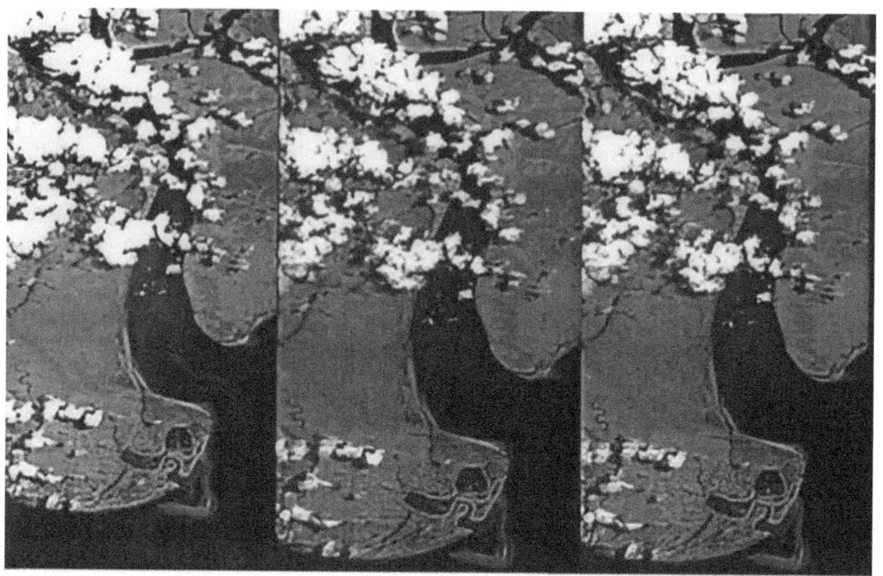

FIGURE 3.4 Hyperspectral image of the Sunderban mangrove at 1030nm: view of the scene in the uncorrected band, FLAASH corrected band, and QUAC corrected band.

3.5.1 ANALYSIS OF ATMOSPHERIC CORRECTION RESULTS

After the image bands are resized to 156, the FLAASH and QUAC models are executed on the image. Factors like scene center location, sensor type, flight date, sensor altitude, average ground elevation of the scene, flight time have been used as input for processing of the radiance data. For more accurate correction, the 'tropical atmospheric model' and 'maritime aerosol model' have been used and water vapor content information is extracted from the hyperspectral water absorption bands.

Figure 3.4 depicts the scene in uncorrected band, band after FLAASH correction and after QUAC.

After the running of FLAASH and QUAC models, the haziness in the image got minimized to certain level and the spectral features are sharpened with increased brightness. Figure 3.5 shows the uncorrected, FLAASH corrected, and QUAC corrected spectral profile of the hyperspectral image of Sunderban mangrove study area. After FLAASH correction of the image, it is observed that in the visible portion of the spectrum, the chlorophyll in the plants absorbs the blue and red wavelengths more strongly than green, producing a characteristic small reflectance peak within the green wavelength range. The reflectance then rises sharply across the boundary between red and near infrared wavelengths, which occurs primarily due to interactions with the internal cellular structure of leaves. Figure 3.5 also shows the same image after QUAC. From the profiles depicted in Figure 3.5, the enhancement in vegetation features after running of both the models is easily discernible. It is observed that the dips present in profile wavelengths 625nm to 750nm and 875nm to 1000nm are greatly reduced after FLAASH atmospheric correction. A steep rise is detected

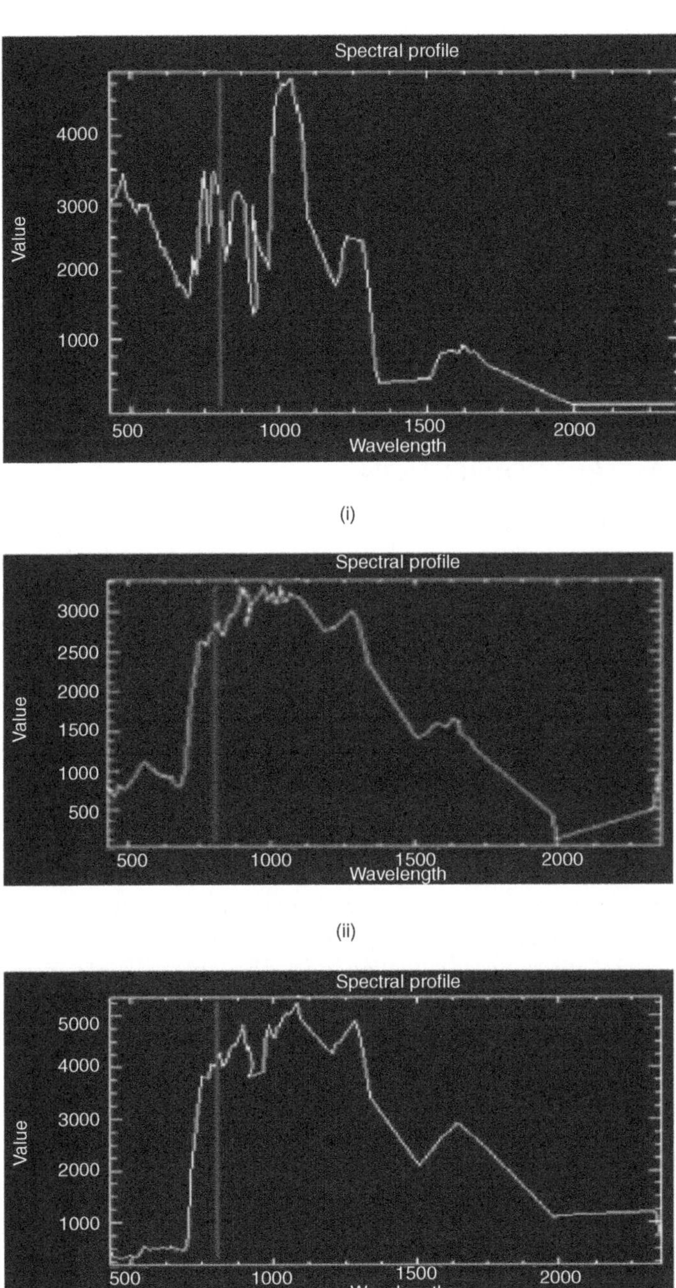

FIGURE 3.5 Spectral profile of a typical pixel extracted from the hyperspectral image of the Sunderban mangrove for uncorrected data, after FLAASH correction, and after QUAC.

from the NIR region in the atmospherically corrected profile. Also, the presence of a number of narrow contiguous peaks is seen in the 750nm to 1500nm range. After application of QUAC model, there has been enhancement in the vegetation features as compared to original uncorrected hyperspectral image. The correction achieved in the NIR region is better in FLAASH than QUAC.

It is important to choose the appropriate atmospheric correction method and apply it correctly to ensure the best results. Radiative transfer models and dark object subtraction are widely used methods for atmospheric correction. Radiative transfer models such as 6S and FLAASH are widely used and considered as robust methods, while the dark object subtraction method is considered as a simple but efficient method for atmospheric correction. QUAC is considered as a fast and efficient method for atmospheric correction and itis useful for cases where computational resources are limited.

3.6 GEOMETRIC CORRECTION

Geometric correction is a preprocessing step that is applied to hyperspectral data to correct for geometric distortions, such as perspective, scale, and rotation. It is a crucial step in making sure that the data is accurately represented spatially, and it can improve the accuracy and quality of the data for further analysis.

1. Image-to-image registration: image-to-image registration is a method that is used to align two or more images of the same scene taken at different times or from different viewpoints. This method uses various feature-based or intensity-based techniques to align the images and correct for geometric distortions.
2. Rational polynomial coefficients (RPC): RPC is a method that uses polynomial functions to model the geometric distortion of the image. This method requires a set of control points, which are used to estimate the polynomial coefficients that correct for the geometric distortion.
3. Sensor modeling: sensor modeling is a method that uses information about the sensor, such as the sensor's position, orientation, and focal length, to model the geometric distortion of the image. This method is commonly used for images acquired from satellite or aircraft-based sensors.

Geometric correction is an important step in preprocessing of hyperspectral data as it corrects for geometric distortions, such as perspective, scale, and rotation. It is a crucial step in making sure that the data is accurately represented spatially, and it can improve the accuracy and quality of the data for further analysis. In a study of mangrove species classification, geometric correction can help to improve the accuracy of the results by removing errors that can cause misclassification of materials or substances in the scene.

3.6.1 IMAGE-TO-IMAGE REGISTRATION

Image-to-image registration is a method that is used to align two or more images of the same scene taken at different times or from different viewpoints. This method uses various feature-based or intensity-based techniques to align the images and correct for geometric distortions.

1. Feature-based registration: feature-based registration is a method that uses image features, such as points, lines, or corners, to align the images. This method is based on the idea that the same features should appear in the same locations in the images. The most common feature-based registration methods are scale-invariant feature transform (SIFT) and speeded up robust features (SURF).

2. Intensity-based registration: intensity-based registration is a method that uses the intensity or gray-level values of the pixels to align the images. This method is based on the idea that the same intensity values should appear in the same locations in the images. The most common intensity-based registration methods are the mutual information and normalized cross-correlation.

Some examples of image-to-image registration are as follows:

- In remote sensing, for example, an image taken by a satellite or aircraft-based sensor may need to be registered with a map or a previous image of the same area. This is to ensure that the information from the image can be accurately overlaid on the map or compared to previous images.
- In medical imaging, image-to-image registration is used to align images from different modalities, such as a CT scan and an MRI, or to align images taken at different times, such as pre- and post-surgery images. This can help to improve the accuracy of the diagnosis and treatment planning.
- In surveillance and security, image-to-image registration is used to align images from different cameras or different viewpoints. This can help to improve the accuracy of object tracking and identification.

Image-to-image registration is an important step in preprocessing of hyperspectral data as it aligns multiple images of the same scene taken at different times or from different viewpoints. This method uses various feature-based or intensity-based techniques to align the images and correct for geometric distortions. It helps to improve the accuracy of the data for further analysis and it is widely used in various applications like remote sensing, medical imaging, and surveillance and security.

3.6.2 SPATIAL INTERPOLATION USING COORDINATE TRANSFORMATIONS

Spatial interpolation using coordinate transformations is a technique used to estimate the value of a pixel in a raster image based on the values of nearby pixels. The technique involves transforming the coordinates of the image from one coordinate system to another, such as from a map projection to a Cartesian coordinate system.

1. Affine transformation: affine transformation is a linear transformation that preserves collinearity and ratios of distances. This method can be used to correct for geometric distortions, such as rotation, translation, and scaling.

2. Polynomial transformation: polynomial transformation is a non-linear transformation that can be used to correct for more complex geometric distortions, such as non-uniform scaling or skewing. This method uses polynomial functions to model the geometric distortion of the image.
3. Radial basis function (RBF) transformation: radial basis function (RBF) transformation is a non-linear transformation that uses a set of basis functions to model the geometric distortion of the image. This method can be used to correct both linear and non-linear geometric distortions.

Spatial interpolation using coordinate transformations is an important step in preprocessing of hyperspectral data as it helps to correct for geometric distortions in the image. It can be used to correct for distortions, such as rotation, translation, scaling, skewing, and non-uniform scaling. With the help of this technique, the image can be registered to a reference image or map projection. This can help to improve the accuracy of the data for further analysis.

3.6.3 Intensity Interpolation and Resampling

Intensity interpolation and resampling are techniques used to improve the spatial resolution or spectral resolution of a hyperspectral image. These techniques are used to estimate the value of a pixel based on the values of nearby pixels.

1. Intensity interpolation: intensity interpolation is a technique used to estimate the value of a pixel based on the importance of nearby pixels. This technique can be used to improve the spatial resolution of an image by estimating the value of a pixel in a higher-resolution image based on the importance of nearby pixels in a lower-resolution image. The most common intensity interpolation methods are nearest-neighbor interpolation, bilinear interpolation, and bicubic interpolation.
2. Resampling: resampling is a technique used to change the resolution of an image. This technique can be used to improve the spectral resolution of an image by estimating the value of a pixel in a higher-resolution image based on the values of nearby pixels in a lower-resolution image. The most common resampling methods are nearest-neighbor resampling, bilinear resampling, and bicubic resampling (Jensen, 1996).

Nearest-Neighbor (NN) Resampling Method
In nearest-neighbor (or zero order) interpolation, the brightness values closest to the x', y' coordinate specified is assigned to the output x, y coordinate. This resampling method is most suited in this application as it does not alter the pixel intensity values during resampling. This method is most appropriate for the mangrove forest image scene because only a subtle change in intensity value makes a lot of difference in discriminating one mangrove species fromanother.

Bilinear Interpolation (BIL) Method
Bilinear (or first-order) interpolation assigns output pixel values by interpolating brightness values in two orthogonal directions in the input image. It basically fits a plane to the four pixel values nearest to the desired point (x', y') in the input image and computes a new brightness value based on weighted distances to these points.

Cubic Convolution (CC) Resampling Method
The cubic convolution method assigns values to output pixels in the same manner as bilinear interpolation except that the weighted values of 16 input pixels surrounding the location of the desired pixel are used to determine the value of the output pixel.

Intensity interpolation and resampling are essential steps in preprocessing hyperspectral data as they help improve an image's spatial and spectral resolution. These techniques can be used to estimate the value of a pixel based on the importance of nearby pixels and can help to improve the accuracy of the data for further analysis.

3.6.4 Accuracy Assessment of Ground Control Points

Accuracy assessment of ground control points (GCPs) is the process of evaluating the accuracy of the location of GCPs in a hyperspectral image. GCPs are known locations on the ground that are used to accurately geo-reference an image, allowing for accurately measuring distances and areas.

1. Checkerboard method: one way to assess the accuracy of GCPs is to use the checkerboard method. This method involves creating a grid of GCPs on the ground, typically using a GPS receiver, and then comparing the locations of these GCPs to the areas of the corresponding pixels in the image.
2   Root mean square error (RMSE): another way to assess the accuracy of GCPs is to calculate the (RMSE between the known location of the GCPs on the ground and the location of the corresponding pixels in the image. The smaller the RMSE, the more accurate the GCPs are considered to be.

Accurately assessing the location of GCPs is important for ensuring the overall accuracy of a hyperspectral image. By using the checkerboard method or calculating the RMSE, it is possible to determine the level of accuracy of the GCPs and make any necessary adjustments. This can help to improve the accuracy of the data for further analysis.

3.6.5 Analysis of Geometric Correction Results

The outputs of FLAASH corrected image and RST-CC geo-rectified image is shown in Figure 3.6. The RST-BIL and RST-NN geo-rectifications are shown in Figure 3.7. The output of polynomial-CC and polynomial-NN geo-rectification is shown in Figure 3.8.

The NN method is computationally more efficient and faster than BIL and CC resampling methods. This procedure does not alter the pixel brightness values during resampling and is therefore, appropriate for thematic data, such as those of the present

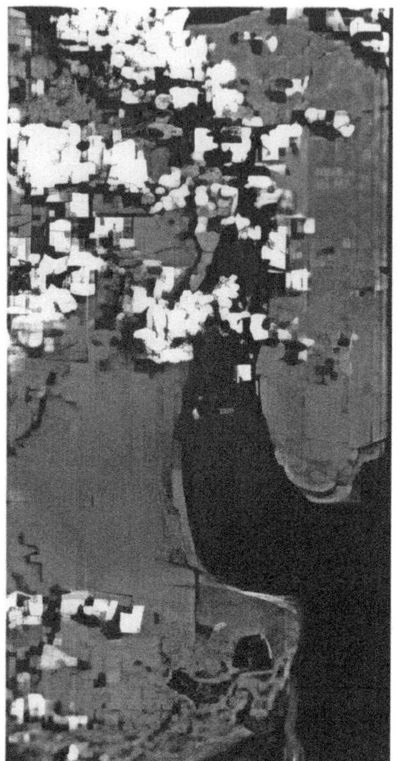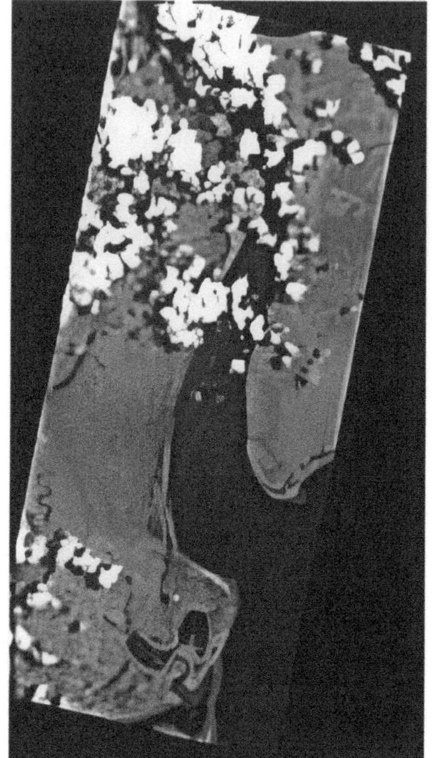

FIGURE 3.6 FLAASH corrected image and RST-CC geo-rectified image.

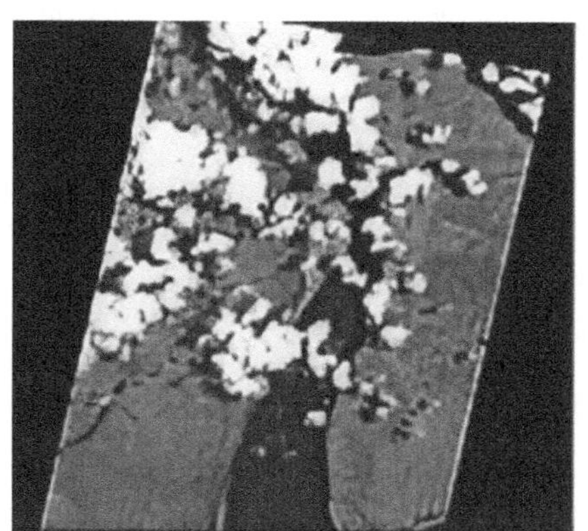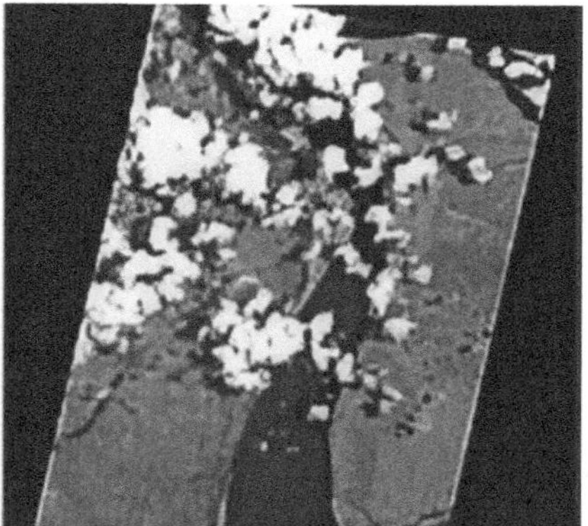

FIGURE 3.7 RST-BIL and RST-NN geo-rectified image.

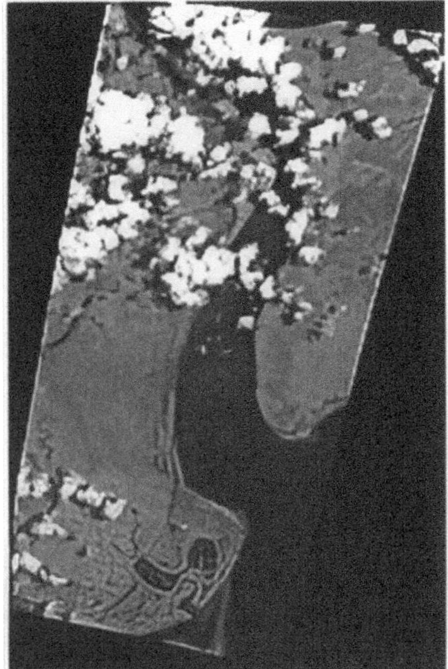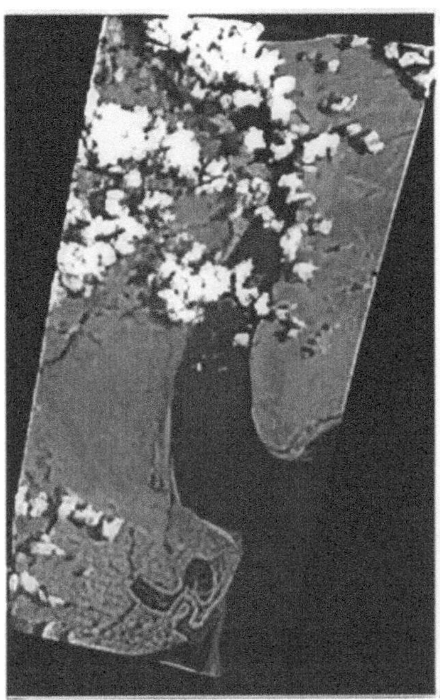

FIGURE 3.8 Polynomial-CC and polynomial-NN geo-rectified image.

study. The only drawback observed is the staircase or jagged effect in NN corrected images. BIL and CC interpolations make use of averages to compute the output intensity, which often leads to removal of valuable spectral information from the images. The BIL method acts as a spatial moving filter that subdues extremes in brightness value throughout the output image. This makes it a computationally demanding method. But the staircase effect observed in the NN approach is greatly reduced and smoothened. The CC corrected image has a dramatic smoothing effect as compared to the NN and BIL methods because of the averaging out of a greater number of pixels. CC and BIL techniques are only appropriate for continuous data such as elevation or slope.

4 Endmember Detection

4.1 AUTOMATED ENDMEMBER DETECTION ALGORITHMS

Automated endmember detection is a crucial step in the processing of hyperspectral data, as it allows for the identification of the pure spectral signatures present within a mixed pixel. The endmember detection process involves identifying the unique spectral signatures of different materials present in an image, also known as endmembers. These endmembers can then be used to determine the abundance of each material in each pixel through a process called spectral unmixing. In this chapter, we will review various automated endmember detection algorithms and their applications in the classification of mangrove species using hyperspectral data. We will explore unsupervised techniques such as N-FINDR, automated target generation process (ATGP), and pixel purity index (PPI). The performance of these algorithms will be evaluated and compared using ground reference data. The most effective algorithm will be selected for use in the subsequent analysis of the hyperspectral data.

This information is valuable for a wide range of applications, including land cover classification, mineral mapping, and vegetation analysis. In this chapter, we will explore various automated endmember detection algorithms and their applications in the context of mangrove species classification. We will delve into the strengths and limitations of each algorithm and analyze the results of their implementation on real-world hyperspectral data. The ultimate goal is to provide a comprehensive understanding of automated endmember detection and its role in hyperspectral image analysis.

In this chapter, we will review these algorithms in detail, analyze their performance, and provide examples of their application to mangrove species classification using hyperspectral data. In the context of this book, automated endmember detection algorithms are crucial for the accurate classification and unmixing of mangrove species. The unique spectral signatures, or endmembers, of different mangrove species, can be extracted using these algorithms, providing a representative set of endmembers for further analysis.

Additionally, endmember detection algorithms can also be used to improve the accuracy of the data by removing noise and other artifacts present in the data, resulting

DOI: 10.1201/9781003432623-4

in a more representative set of endmembers. This is particularly important in the case of mangrove species classification, as the spectral signatures of different species can be very similar and difficult to differentiate.

Furthermore, automated endmember detection algorithms can also help reduce the data's dimensionality, making it more manageable for further analysis and classification. This is particularly important in the case of mangrove species classification, as the data can be pretty significant and complex, making it difficult to process and analyze.

The chapter focuses on three important algorithms for endmember detection.

4.1.1 AUTOMATED TARGET GENERATION PROCESS (ATGP)

ATGP is an algorithm that is used to automatically detect endmembers (pure spectral signatures) in hyperspectral data. The algorithm works by first creating a set of candidate endmembers using a variety of techniques, such as principal component analysis (PCA) and independent component analysis (ICA). These candidates are then evaluated against the data using various metrics, such as the angle between the candidate and the data, the volume of the convex hull formed by the candidate and the data, and the residual sum of squares (RSS) between the candidate and the data. The candidates that have the highest scores are then selected as the final endmembers.

The ATGP algorithm can be broken down into several steps:

1. Data preprocessing: the first step in the ATGP algorithm is to preprocess the data by removing noise, atmospheric effects, and geometric distortions.
2. Candidate generation: the next step is to generate a set of candidate endmembers. This can be done using techniques such as PCA, ICA, or other techniques that can identify distinct spectral signatures in the data.
3. Evaluation of candidates: the candidates are then evaluated against the data using various metrics, such as angle, the volume of the convex hull, and RSS.
4. Selection of endmembers: the candidates that have the highest scores are selected as the endmembers.
5. Post-processing: the final step is to post-process the endmembers and data to remove any remaining noise or artifacts.

The ATGP algorithm has been found to be effective in detecting endmembers in hyperspectral data, particularly for images with high spectral dimensionality. However, the algorithm can be sensitive to the initial conditions and the choice of clustering technique used.

An example of using the ATGP algorithm on a hyperspectral dataset of a coastal area, the algorithm is applied to the preprocessed data, and the resulting endmembers are used to create a spectral library that can be used for classification and mapping of the different materials present in the area.

4.1.2 N-FINDER

The N-FINDR algorithm is a widely used automated endmember detection algorithm for hyperspectral data. The algorithm is based on a non-parametric method that uses statistical analysis to identify the endmembers within a hyperspectral data set.

The N-FINDR algorithm works by first performing a principal component analysis (PCA) on the data set to reduce the dimensionality of the data. This is done to ensure that the algorithm only considers the most important features of the data. Once the PCA is performed, the algorithm then uses a statistical method called the minimum volume enclosing ellipsoid (MVEE) to identify the endmembers.

The MVEE method is based on the idea that the endmembers in a data set should be the most distinct and well-separated group of points within the data. To accomplish this, the algorithm uses the PCA-reduced data to calculate the covariance matrix of the data, and from this matrix, it calculates the eigenvectors and eigenvalues. These eigenvectors and eigenvalues are then used to construct an ellipsoid that encloses the data, to find the minimum volume ellipsoid that encloses all of the endmembers.

Once the MVEE is calculated, the algorithm uses a statistical method called the Mahalanobis distance to determine which data sets are the most distinct and well-separated points. These points are considered to be the endmembers.

The mathematical explanation of the N-FINDR algorithm would involve matrix algebra, eigenvalue decomposition, and multivariate statistical analysis.

4.1.3 PIXEL PURITY INDEX (PPI)

PPI is an automated endmember detection algorithm that is based on the concept of pixels being pure or impure. The algorithm uses the spectral information of a pixel to determine its purity, and it does this by comparing the pixel's spectral signature to the spectral signatures of the endmembers. PPI is based on the assumption that pure pixels have spectral signatures that are similar to the endmembers, while impure pixels have spectral signatures that are a linear combination of the endmembers.

The algorithm works by first initializing the endmembers using a set of randomly selected pixels. Then, for each pixel in the image, the algorithm calculates the PPI value using the following formula:

$$PPI = (1\ /\ (1 + d))$$

where d is the Euclidean distance between the pixel's spectral signature and the closest endmember's spectral signature. The pixels with the highest PPI values are considered to be the purest pixels and are used to update the endmembers. The process is repeated until the endmembers converge or a maximum number of iterations is reached.

One of the advantages of PPI is that it is relatively simple to implement and computationally efficient. It also does not require any prior knowledge about the endmembers or the number of endmembers present in the image. However, one of the limitations of PPI is that it can be sensitive to noise and outliers in the data, which can lead to inaccurate endmember estimates.

4.2 SPECTRAL SIGNATURE ANALYSIS OF EXTRACTED ENDMEMBERS

After the endmember detection process, it is essential to analyze the spectral signatures of the extracted endmembers to ensure that they are pure and distinct. Spectral signature analysis can be done using various techniques, such as scatter plots, angle-based similarity measures, and Mahalanobis distance.

Scatter plots can be used to visualize the similarity between two endmembers. The scatter plot of two endmembers is created by plotting the reflectance values of the two endmembers at each wavelength. If the two endmembers are pure, their scatter plot will be a straight line. If the two endmembers are not pure, their scatter plot will be a cloud of points.

Angle-based similarity measures can be used to quantitatively measure the similarity between two endmembers. The angle-based similarity measures are based on the angle between the two endmembers in the reflectance space. The smaller the angle, the more similar the two endmembers are.

Mahalanobis distance is a measure of distance between two endmembers in the reflectance space. The Mahalanobis distance is based on the covariance matrix of the reflectance data. The smaller the Mahalanobis distance, the more similar the two endmembers are.

When analyzing the spectral signature of the extracted endmembers using N-FINDR, one would look for patterns such as smoothness, uniqueness, and linearity. The smoothness of the spectral signature indicates that the endmember is well-defined and not affected by noise. The uniqueness of the spectral signature indicates that the endmember is not a linear combination of other endmembers. The linearity of the spectral signature indicates that the endmember is a pure pixel.

When analyzing the spectral signature of the extracted endmembers using ATGP, one would look for patterns such as smoothness, uniqueness, and high similarity to the target. The smoothness of the spectral signature indicates that the endmember is well-defined and not affected by noise. The uniqueness of the spectral signature indicates that the endmember is not a linear combination of other endmembers. High similarity to the target indicates that the endmember is a pure pixel.

When analyzing the spectral signature of the extracted endmembers using PPI, one would look for patterns, such as smoothness, uniqueness, and high pixel purity. The smoothness of the spectral signature indicates that the endmember is well-defined and not affected by noise. The uniqueness of the spectral signature indicates that the endmember is not a linear combination of other endmembers. High pixel purity indicates that the endmember is a pure pixel.

It is important to note that these are just examples of a few spectral signature analysis techniques, there are many other techniques as well, which can be used for this purpose depending on the requirement and dataset.

4.3 EXPERIMENTAL RESULTS OF ANALYSIS BETWEEN ENDMEMBER DETECTION ALGORITHMS

Ground survey of Henry Island indicates the dominance of seven mangrove species in the area. The ATGP and N-FINDR algorithms have been set to extract seven pure endmember signatures of mangrove species from the hyperspectral image scene. On application of automated target detection algorithms, the unknown spectral signatures of target endmembers have been extracted and identified (Plaza et al., 2004). The locations of pure spectra and extracted spectral signatures of mangrove species identified by the N-FINDR algorithm are shown in Figure 4.1 and Figure 4.2, respectively. Likewise, Figures 4.3–4.10 display the spectral signatures and respective endmember locations for different values of p(no. of endmembers), extracted by the ATGP algorithm.

To evaluate the performance of the two algorithms, the input number of endmembers (p) has been varied from 7 to 19 to check the efficiency of the algorithms for detecting the correct signature of mangrove species (endmembers). Table 4.1 shows the number of pure mangrove species extracted by N-FINDR and ATGP for varying values of endmembers. In the present study, N-FINDR has accurately identified seven target endmembers at $p = 7$, whereas ATGP identified the same number at $p = 10$. Figures 4.3–4.10 display the spectral signatures of these endmembers (extracted by ATGP) and their respective locations at different values of p. An algorithm is considered better if it can extract different endmember signatures (here the maximum extracted is 7) at a smaller value of p, which means it can efficiently cull

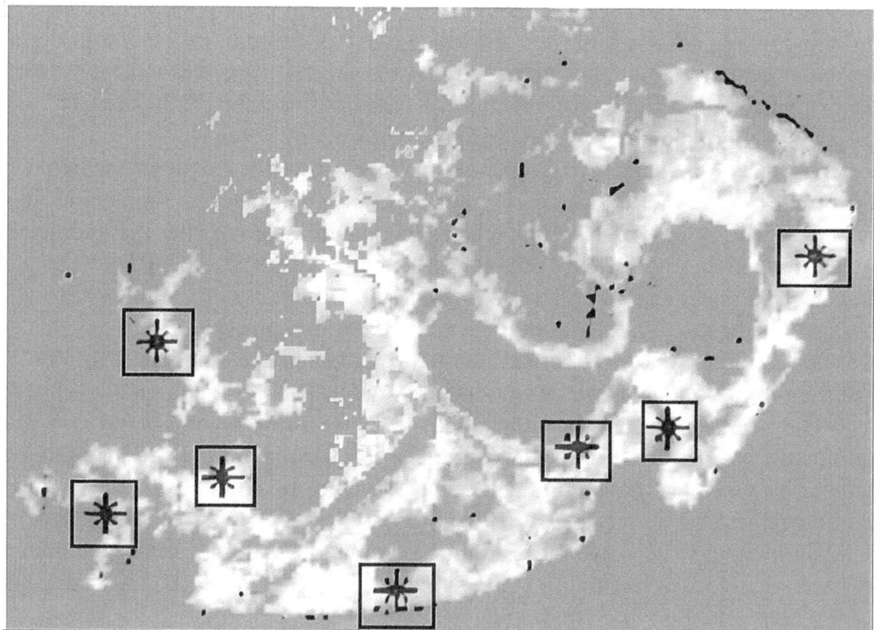

FIGURE 4.1 Locations of endmembers extracted using N-FINDR.

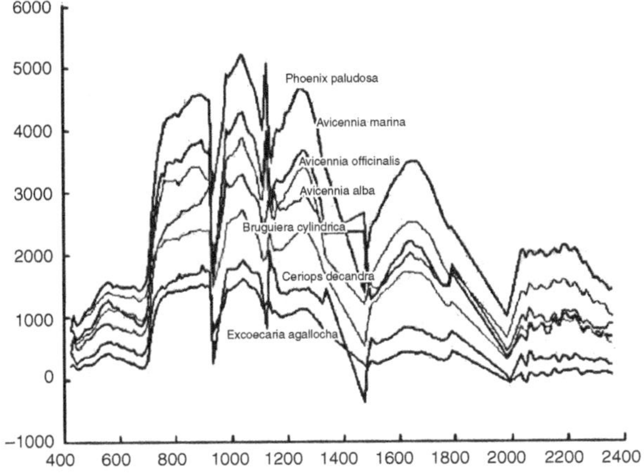

Excoecaria agallocha *Ceriops decandra* *Phoenix paludosa* *Avicennia alba*

Avicennia marina *Bruguiera cylindrica* *Avicennia officinalis*

FIGURE 4.2 Spectral profile of mangrove species generated by the N-FINDR algorithm.

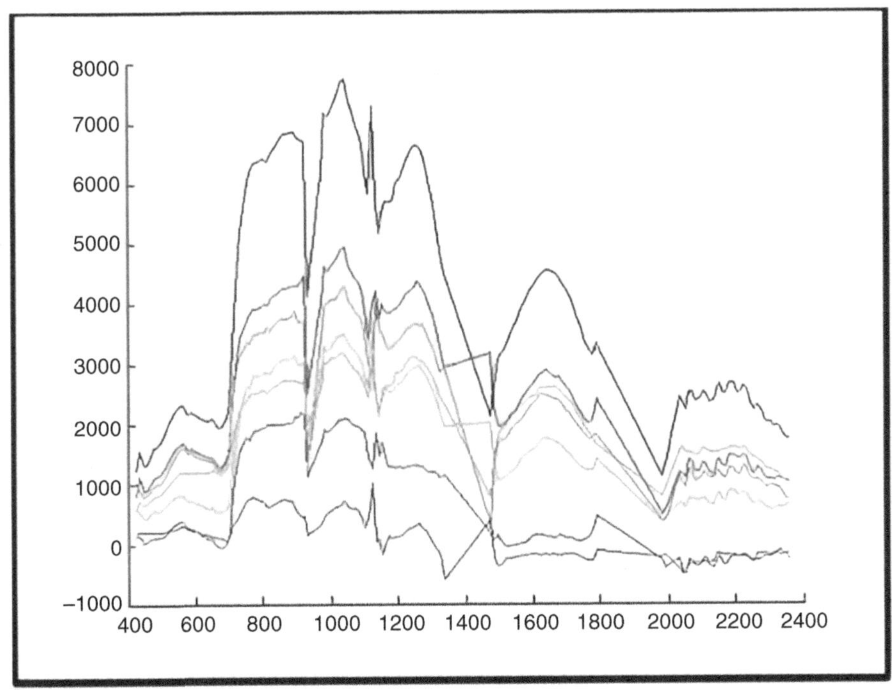

FIGURE 4.3 ATGP endmembers with $p=7$.

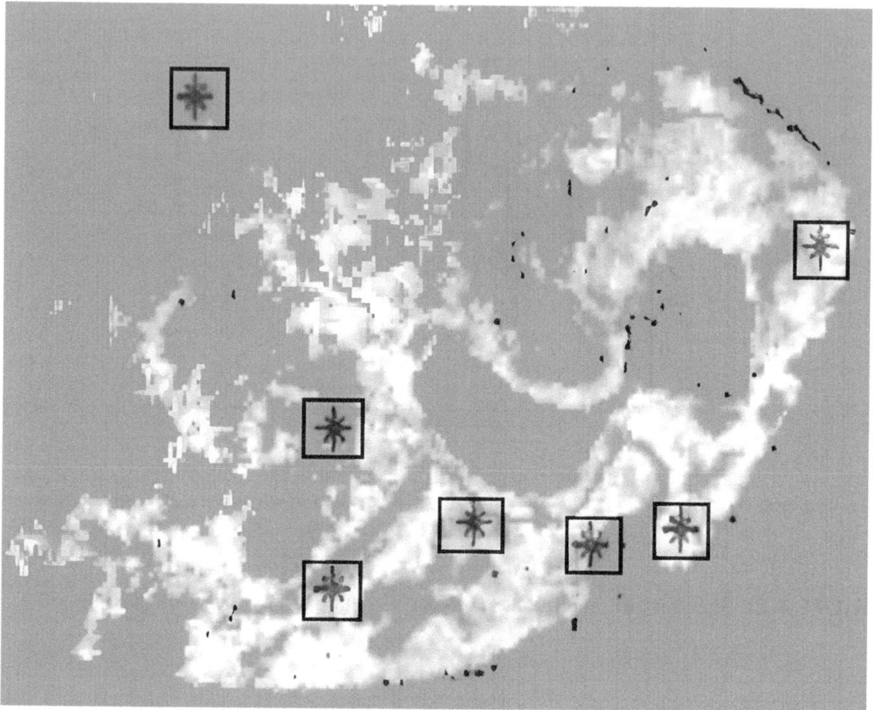

FIGURE 4.4   Location of seven endmembers in the study area.

distinctive pixels. Taking this factor into account, N-FINDR is found here to be more efficient than ATGP. As ATGP detects the most dissimilar pixel from a set of similar pixels (of vegetation), it has not been able to detect pure pixels (endmembers) with much accuracy compared to N-FINDR. N-FINDR has estimated endmembers by finding out pixel sets with largest possible volume by 'inflating' a simplex inside the data. This has been very well determined from the hyperspectral image of the study area.

The PPI algorithm has been executed with 10,000 iterations and could extract six distinct pure pixels (mangrove species) from the forest image scene as compared to seven pure pixels identified by ATGP and N-FINDR. PPI has not been able to identify *Phoenix paludosa*, which N-FINDR and ATGP has done successfully. Figures 4.11 and 4.12 display the locations of endmembers identified by PPI and their spectral profiles, respectively. A threshold value of 900 has been fixed in this study to designate a pure pixel. The six pixels that have been identified as pure have counts of 2012, 1226, 1107, 1000, 999, and 976. The pixels below 900 count display spectral profiles that are either similar to the profiles of higher count or do not resemble vegetation profiles. Hence they have been discarded.

The endmember signatures collected during ground survey has been plotted on the hyperspectral image of the study area (Figure 4.13). These signatures have been considered as reference signatures for the image extracted mangrove species. The

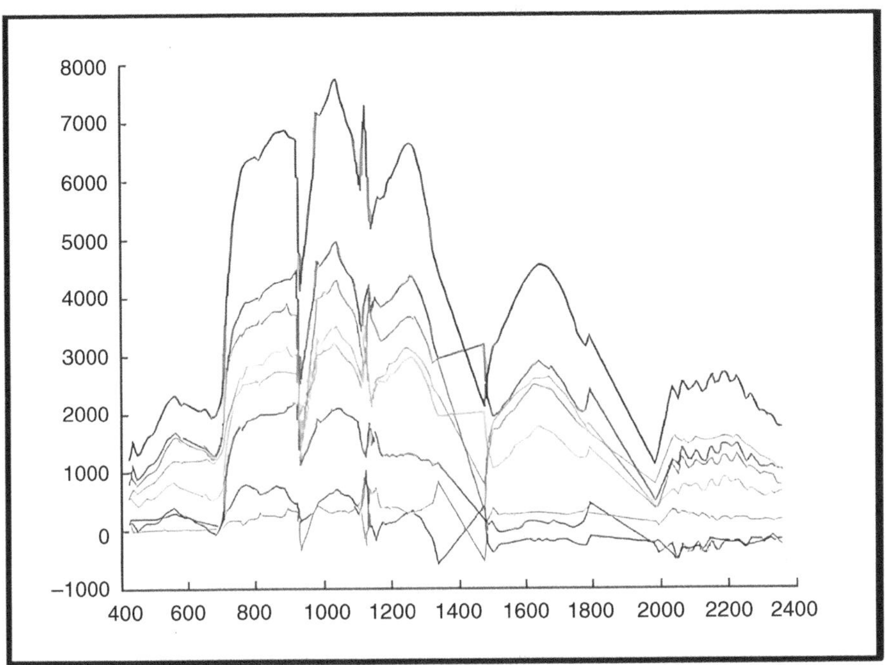

FIGURE 4.5 ATGP endmembers with $p=8$.

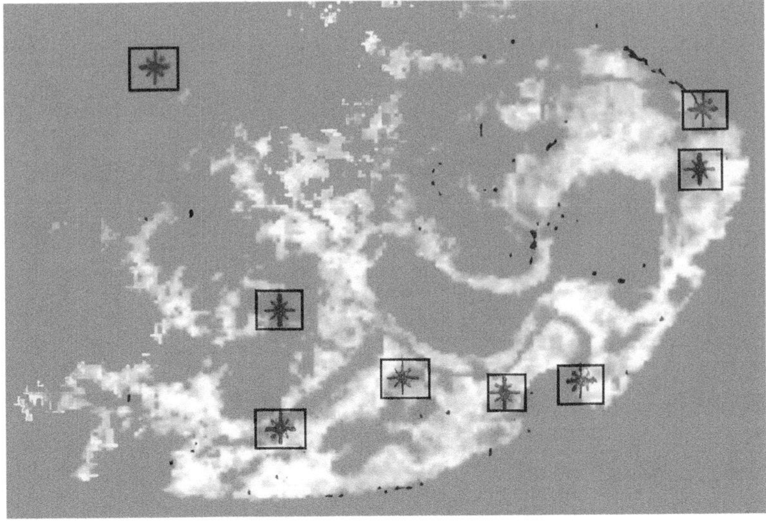

FIGURE 4.6 Location of eight endmembers in the study area.

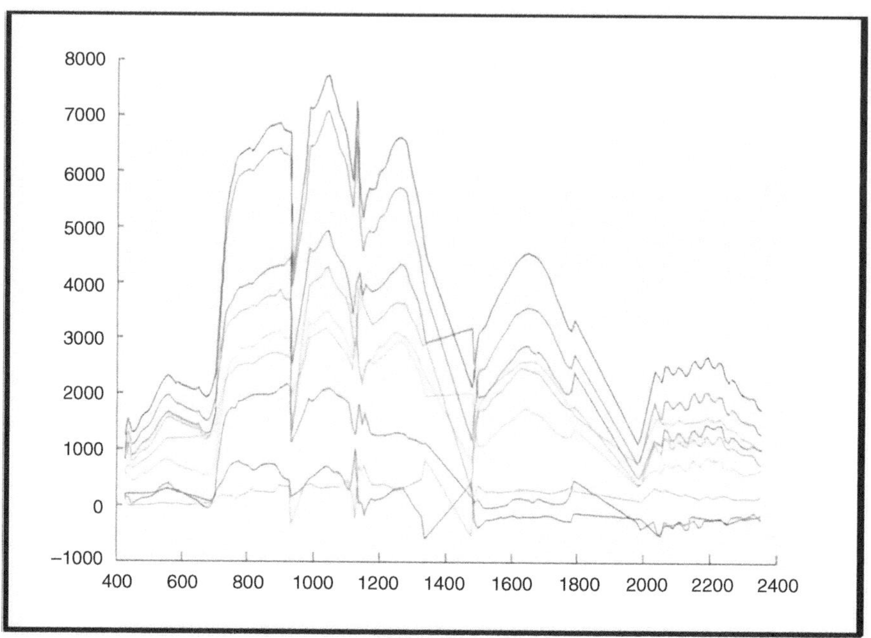

FIGURE 4.7   ATGP endmembers with p=9.

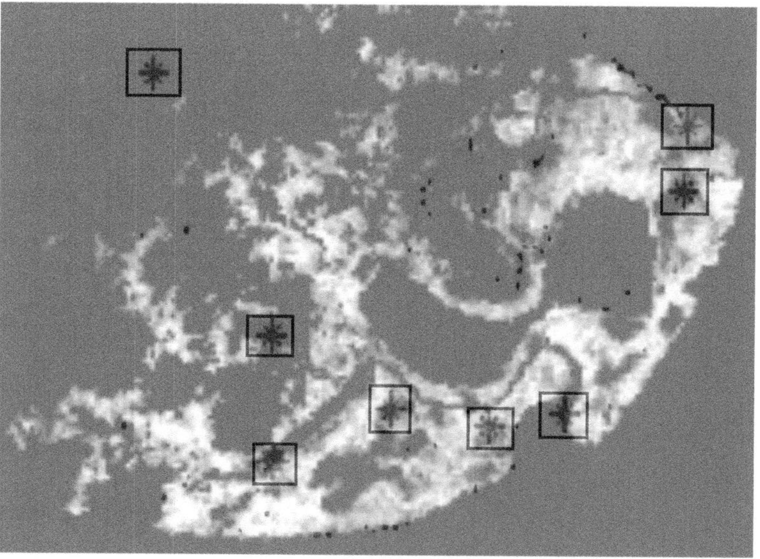

FIGURE 4.8   Location of nine endmembers in the study area.

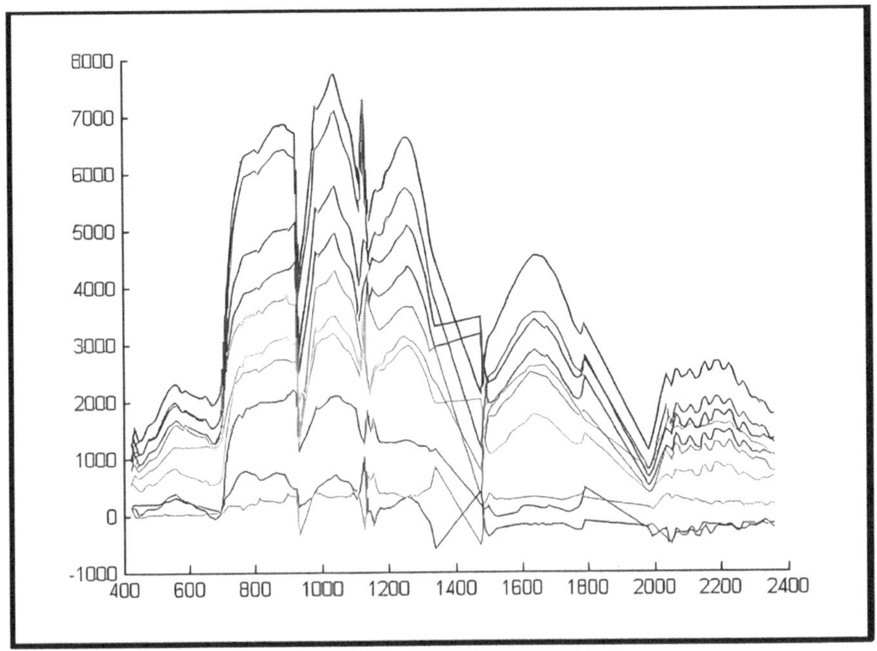

FIGURE 4.9 ATGP endmembers with p=10.

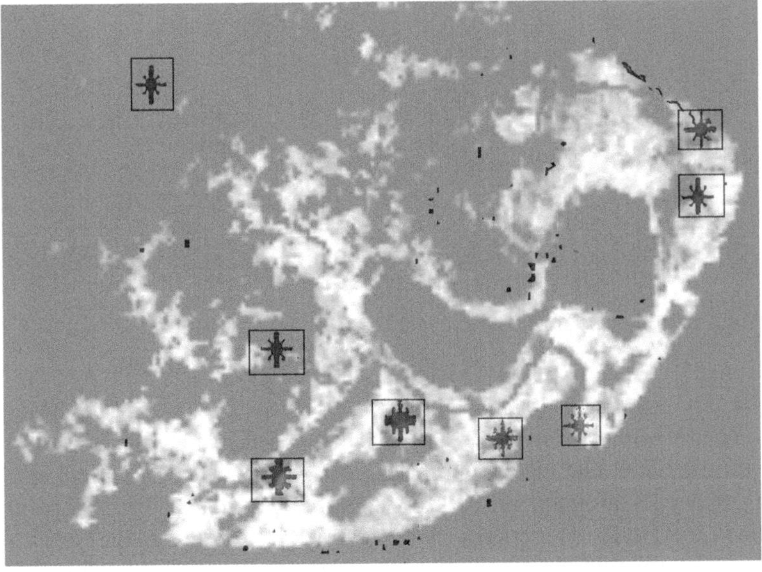

FIGURE 4.10 Location of ten endmembers in the study area.

TABLE 4.1

Number of Pure Mangrove Species Extracted by N-FINDR and ATGP for Various Values of *p*

Endmember Detection	Number of Endmembers (*p*)												
Algorithms Applied	7	8	9	10	11	12	13	14	15	16	17	18	19
N-FINDR	7	7	7	7	7	7	7	7	7	7	7	7	7
ATGP	5	5	6	7	7	7	7	7	7	7	7	7	7

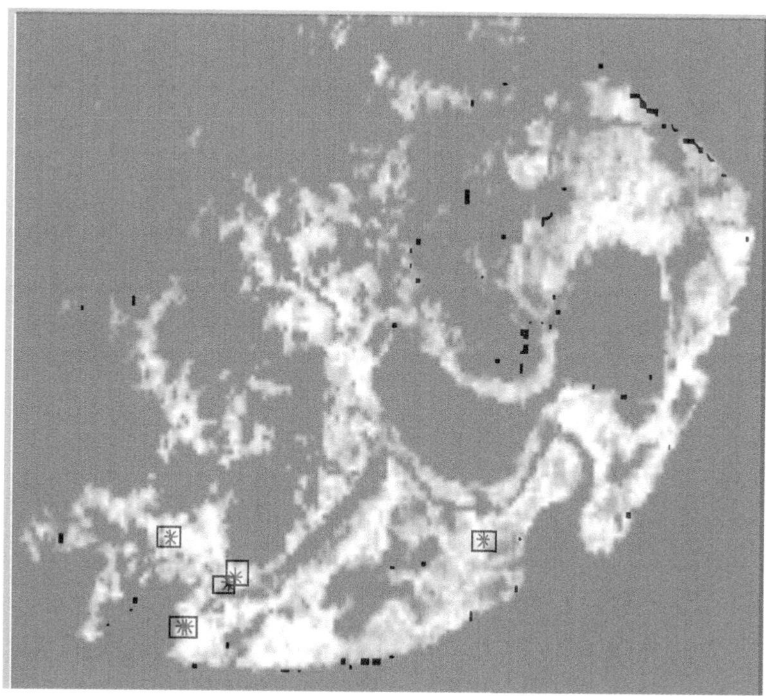

FIGURE 4.11   Plots of pure pixels extracted by PPI.

endmember signatures extracted with ATGP, N-FINDR, and PPI algorithms have been compared with the reference spectra using spectral angle mapper (SAM) algorithm, an automated method for comparing image spectra to individual spectra or a spectral library (Boardman, 1995; Kruse et al., 2000; Jensen, 1996). The algorithm determines the similarity between two spectra by calculating the 'spectral angle' between them, treating them as vectors in a space with dimensionality equal to the

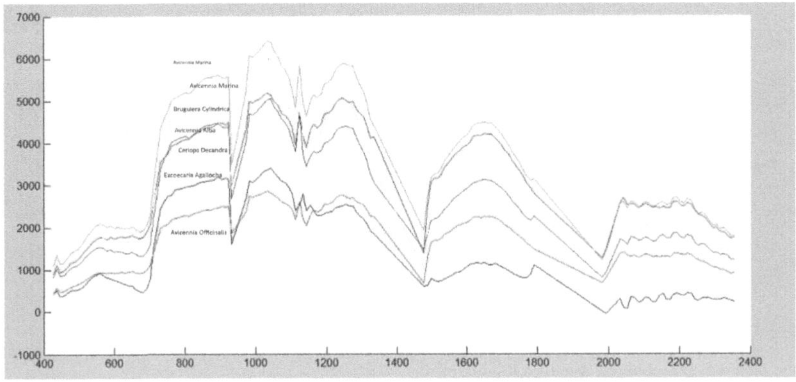

Excoecaria agallocha Avicennia alba

Ceriops decandra   *Avicennia marina*

Avicennia officinalis   *Bruguiera cylindrica*

FIGURE 4.12   Endmembers identified with PPI algorithm.

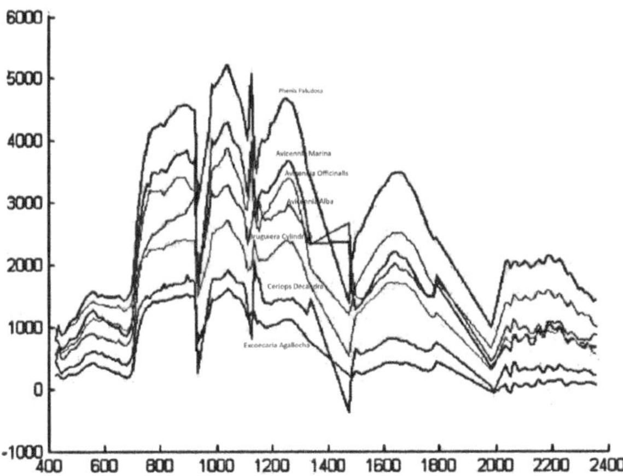

Excoecaria agallocha   Ceriops decandra   Phoenix paludosa   Avicennia alba

Avicennia marina   Bruguiera cylindrica   Avicennia officinalis

FIGURE 4.13   Spectral library of seven dominant mangrove species plotted during the ground survey in Henry Island.

TABLE 4.2

Spectral Angle between Mangrove Signatures (Ground Truth) and Corresponding Extracted Signatures

Endmember Detection Algorithm Applied	Endmembers (Species) Identified						
	Ceriops decandra	Avicennia marina	Avicennia alba	Avicennia officinalis	Excoecaria agallocha	Bruguiera cylindrica	Phoenix paludosa
ATGP	1.910	3.020	2.840	2.640	2.260	2.530	2.140
N-FINDR	2.880	1.380	2.170	1.470	1.970	1.970	1.880
PPI	2.99	2.55	2.65	2.10	1.15	2.10	Not Identified

number of bands (*nb*). SAM determines the similarity of an unknown spectrum *t* to a reference spectrum *r*, by applying the following equation:

$$\alpha = \cos^{-1}\left(\frac{\sum\limits_{i=1}^{nb} t_i r_i}{\left(\sum\limits_{i=1}^{nb} t_i^2\right)^{1/2}\left(\sum\limits_{i=1}^{nb} r_i^2\right)} \right)$$

where '*nb*' equals the number of bands in the image.

For each reference spectrum chosen in the analysis of a hyperspectral image, the spectral angle is determined for every image spectrum (pixel).

The smaller the average spectral angle the better is the performance, i.e., the signatures identified more closely resemble the actual ground identified signatures (Table 4.2).

On comparison, it is found that except for *Excoecaria agallocha* in PPI, N-FINDR extracted pure mangrove species more closely resemble ground spectra and shows maximum efficiency when compared to PPI and ATGP.

Thus, the application of these algorithms on hyperspectral imagery of the mangrove forested scene has led to the extraction and identification of pure spectra of mangrove species. The spectral library generated by the N-FINDR algorithm has been used for further processing.

5 Least-Squares-Based Linear Spectral Unmixing For Pure Endmembers

5.1 LINEAR SPECTRAL UNMIXING

Linear spectral unmixing (LSU) is a standard method used for the analysis of hyperspectral data. It is based on the assumption that a mixed pixel can be represented as a linear combination of pure endmembers. The least-squares-based LSU method is a popular technique used for the estimation of the fractional abundance of each endmember in a mixed pixel. The method is based on the minimization of the residuals between the observed mixed pixel and the estimated mixed pixel, which is a linear combination of the endmembers. The residuals are calculated as the difference between the observed mixed pixel and the estimated mixed pixel. The fractional abundance of each endmember in a mixed pixel is estimated by minimizing the residuals.

The LSU method can be applied to both pure and mixed pixels. For pure pixels, the fractional abundance is set to 1 for the corresponding endmember and 0 for all other endmembers. For mixed pixels, the fractional abundance is estimated using the least-squares method. The LSU method is sensitive to the presence of noise and can produce inaccurate results when the noise is high. To overcome this problem, various techniques, such as the constrained least squares, the total least squares, and the orthogonal least squares, have been proposed.

The LSU method has been widely used for the analysis of hyperspectral data in various applications, such as mineral mapping, vegetation mapping, and urban land use mapping. The method has been found to be effective in the estimation of the fractional abundance of endmembers in mixed pixels. However, the method assumes that the endmembers are pure and that the mixed pixels can be represented as a linear combination of the endmembers. This assumption is not always satisfied in practice, and various non-linear unmixing methods have been proposed to overcome this limitation.

The LSU model is based on the principle that if the total surface area is divided proportionally based on fractional abundances of constituent entities, the reflected radiation bears the characteristics of these entities in same proportion (Keshava and Mustard, 2002). The relationship envisaged between the fractional abundance of entities comprising image and the spectrum of reflected radiation is linear.

DOI: 10.1201/9781003432623-5"

The algorithm is described as follows.

Initialization: let t_1, t_2, t_3,..., t_p represent p targets and L, the number of spectral bands. Let m_1, m_2,..., m_p denote their respective spectral signature vectors, which are generally their reflectance values in each band.

A linear spectral mixture of r is a linear combination of m_1, m_2,..., m_p with appropriate abundance fractions α_1, α_2, α_3,..., α_p. Here r and M is an $L \times 1$ column pixel vector and an $L \times p$ target spectral signature matrix, denoted by $[m_1, m_2,..., m_p]$, respectively. m_j is an $L \times 1$ column vector for $1 \leq j \leq p$. Also let $\alpha = (\alpha_1, \alpha_2, \alpha_3,..., \alpha_p)T$ be a $p \times 1$ abundance column vector associated with r and α_j denoting the abundance fraction of the jth target signature m_j present in the pixel vector r.

The spectral signature of an image pixel r is modeled as

$$r = M\alpha + n$$

where n is noise or can be interpreted as model error.

In the above algorithm, the t_1, t_2, t_3,..., t_p endmembers and m_1, m_2,...,m_p spectral signature vectors have been identified using automated target detection algorithms (discussed in details in Chapter 4). LSU is an unconstrained model that calculates the fractional abundance values $(\alpha_1, \alpha_2, \alpha_3,..., \alpha_p)$ of the endmembers in the pixel vector 'r'.

Two constraints have been imposed on this model to arrive at the desired solution. These are

Abundance sum-to-one-constraint (ASC) that is $\Sigma p_j = 1$ $\alpha_j = 1$: The sum of all abundances is equal to one.

Abundance non-negativity constraint (ANC) that is $\alpha_j > 0$ for all $1 \leq j \leq p$. The abundance values cannot be less than 0.

The LSU model with the above two constraints is known as a fully constrained linear spectral unmixing model.

Fully constrained LSU is used for quantification of endmembers present in pixel vector 'r' wherein their true abundance fractions can be estimated accurately. Such endmember quantification is not achieved in the unconstrained LSU method. However, only for target detection studies, exact estimation of target abundance fraction may not be essential. As long as estimated abundances of the target pixel vectors can distinguish themselves from surrounding pixel vectors, the targets can be detected effectively even if the abundance fractions of LSU do not satisfy ASC or ANC. To implement full constraints (ASC and ANC), the target abundance fractions must be constrained to the range of [0, 1].

There are several least-squares-based linear spectral unmixing algorithms, including the fully constrained least squares (FCLS), the minimum volume constraint algorithm (MVCA), and the vertex component analysis (VCA). These algorithms differ in the way they handle the constraints imposed by the LMM, but they all share the goal of finding the set of endmembers and their fractions that best fit the observed data.

The FCLS algorithm is the most widely used least-squares-based linear spectral unmixing algorithm. It finds the set of endmembers and their fractions that minimize the reconstruction error, subject to the constraint that the fractions must be non-negative and sum to one. This algorithm is simple to implement and fast, but it is sensitive to noise and outliers in the data.

The MVCA algorithm is a variation of the FCLS algorithm that handles the non-negativity and sum-to-one constraints differently. It finds the endmembers and their fractions that minimize the volume of the convex hull of the data, subject to the constraint that the fractions must be non-negative. This algorithm is more robust to noise and outliers than the FCLS algorithm, but it is more computationally intensive.

The VCA algorithm is a variation of the FCLS algorithm that finds the endmembers and their fractions by identifying the vertices of the convex hull of the data. This algorithm is less sensitive to noise and outliers than the FCLS algorithm, but it is also more computationally intensive.

In the least-squares-based linear spectral unmixing (LSU) method, the non-negative and abundance sum-to-one constraints are imposed on the abundance maps. These constraints ensure that the estimated abundances are physically meaningful, i.e., they must be non-negative and the sum of all abundances for each pixel must be equal to one.

The non-negative constraint is imposed to ensure that the estimated abundances are greater than or equal to zero, as negative abundances do not make physical sense. The abundance sum-to-one constraint is imposed to ensure that the total fraction of each pixel is accounted for by the different endmembers. This means that the sum of all abundances for each pixel must be equal to one.

These constraints are important in the LSU method as they help to ensure that the estimated abundances are physically meaningful, and they also help to improve the stability of the algorithm. However, it should be noted that enforcing these constraints can also lead to some limitations in the method, such as a loss of degrees of freedom and a decrease in the signal-to-noise ratio.

We have discussed the implementation of these constraints in the LSU method and their effect on the results of mangrove species classification using hyperspectral data. We have also compared the performance of the LSU method with and without these constraints and discuss the trade-offs between the two.

In this chapter, we have given a detailed review of the least-squares-based LSU method for the analysis of hyperspectral data. We have explained the advantages and limitations, and its application to real-world datasets. We have also compared the performance of the LSU method with other non-linear unmixing methods in future chapters. Furthermore, we will also provide examples of how this method can be applied to a specific application, such as mangrove species classification, using hyperspectral data.

5.2 LINEAR SPECTRAL UNMIXING OF ENDMEMBERS: CASE STUDY OF MANGROVE ECOSYSTEM

In this section, we will delve deeper into the application of the least-squares-based LSU method for the analysis of hyperspectral data in the context of mangrove ecosystem. We will provide a case study of how this method can be applied to a real-world dataset of mangrove species, highlighting the advantages and limitations of the method in this specific context. We will apply the LSU method to a case study of mangrove ecosystem using a Hyperion dataset of Henry Island in the Sunderbans. The Hyperion sensor, which is operated by NASA's Earth Observing-1 (EO-1) satellite, provides high-resolution hyperspectral data with 224 spectral bands in the visible to shortwave infrared region. The dataset covers an area of approximately 150 square kilometers and was acquired in August 2011. We will also show how the LSU method can be used to extract the spectral signatures of different mangrove species, and how these signatures can be used for accurate classification and mapping of the mangrove ecosystem.

LSU has been attempted by some researchers on multispectral imagery. But their application on hyperspectral image datasets is very recent. Moreover, the use of linear spectral unmixing models for discrimination of pure and mixed mangrove species on hyperspectral data has been rarely studied for the Indian mangrove forests, in general, and the Sunderbans, in particular. Although some work has been done in this direction at the Indian Space Research Organization (ISRO), the same mainly relates to geological and forest canopy studies.

First, we have performed atmospheric and geometric correction on the dataset using standard techniques, such as FLAASH and QUAC. Next, we have extracted the endmembers using automated endmember detection algorithms, such as ATGP, PPI, and N-FINDR. These endmembers will then be used as input for the LSU method, which will be applied to the corrected dataset to estimate the fractional abundance of each endmember in each pixel.

The results of the LSU method have been compared with other linear and non-linear unmixing methods, such as vertex component analysis (VCA) and non-negative matrix factorization (NMF). We have also performed an accuracy assessment of the results using ground control points and confusion matrix analysis.

Furthermore, we have used the fractional abundance maps obtained from the LSU method to classify the mangrove species present in the Henry Island ecosystem. By comparing the classification results with ground truth data, we have evaluated the effectiveness of the LSU method in this application.

Overall, this case study demonstrates the potential of the LSU method for the analysis of hyperspectral data in the context of mangrove species classification and highlight its advantages and limitations. Furthermore, we have also compared the performance of the LSU method with other linear and non-linear unmixing methods in the context of mangrove species classification in further chapters, and discuss the implications of the results for future research in this area.

5.3   CASE STUDY OF FRACTIONAL ABUNDANCE ESTIMATION OF MANGROVE SPECIES

The study of mangrove ecosystems using hyperspectral data can benefit greatly from the use of LSU techniques. In this section, we will present a case study of the application of LSU on a Hyperion dataset of Henry Island in the Sunderbans mangrove forest. The dataset was collected by the NASA EO-1 satellite in 2002 and covers an area of approximately 20 square kilometers. The study area is characterized by a diverse range of mangrove species, including *Rhizophora* spp., *Avicennia* spp., and *Bruguiera* spp.

The first step in the case study was to preprocess the Hyperion dataset. This included atmospheric and geometric correction, removal of bad columns and vertical stripes, and the selection of relevant spectral bands. The next step was to extract endmembers from the preprocessed dataset using automated endmember detection algorithms, such as ATGP and N-FINDR. These endmembers were then used as input in the LSU algorithm for fractional abundance estimation.

We applied both constrained and unconstrained LSU on the dataset. Constrained LSU is a method that applies non-negative and abundance-sum-to-one constraints to the abundance estimates. These constraints ensure that the estimated abundances are physically meaningful. On the other hand, unconstrained LSU does not apply any constraints on the estimated abundances.

The results of the case study showed that the constrained LSU method provided more accurate fractional abundance estimates compared to the unconstrained method. This was evident in the comparison of the estimated fractions with the reference fractions obtained from field measurements. The results also showed that the LSU method performed better than other linear and non-linear unmixing methods, such as the fully constrained least-squares (FCLS) and vertex component analysis (VCA) methods.

Overall, the case study demonstrates the potential of LSU for the accurate estimation of fractional abundances of mangrove species using hyperspectral data. The results of this study can be used to support mangrove conservation and management efforts in the Sunderbans and other similar ecosystems.

5.3.1 ENDMEMBER DETECTION

Endmember detection using algorithms PPI, N-FINDR, and ATGP is discussed in Chapter 4. The mangrove species of the island are considered endmembers and have been automatically detected using the algorithms mentioned. LSU is implemented for sub-pixel classification with the identified endmembers as input on the hyperspectral image having moderate spatial resolution (30m). Figure 5.1 displays the integrated abundance distribution image for mangrove species for unconstrained LSU with N-FINDR endmembers as the input.

5.3.2 FRACTIONAL ABUNDANCE

Fractional abundance has been estimated for endmembers identified by ATGP and PPI algorithms as well. The comparison of fractional abundance of individual mangrove species namely, *Excoecaria agallocha, Ceriops decandra, Phoenix paludosa, Avicennia alba, Avicennia marina, Bruguiera cylindrica, Bruguiera cylindrica, Avicennia officinalis* derived by N-FINDR and ATGP are displayed in Figures 5.2–5.8, respectively. Figure 5.9 displays the classified abundance distribution image with PPI extracted endmembers as input. The classified output (unconstrained and fully constrained) generated with endmembers of N-FINDR show more accuracy with

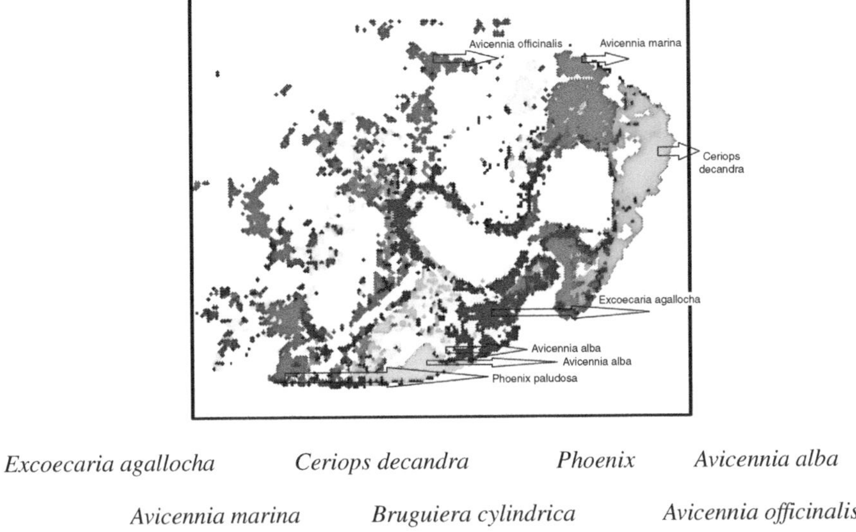

Excoecaria agallocha   *Ceriops decandra*   *Phoenix*   *Avicennia alba*

Avicennia marina   *Bruguiera cylindrica*   *Avicennia officinalis*

FIGURE 5.1   Unconstrained LSU classified image with N-FINDR endmembers.

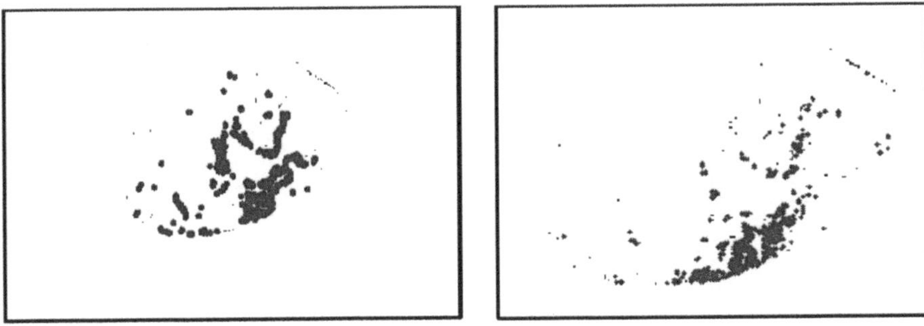

FIGURE 5.2   Unconstrained LSU fractional abundance of *Excoecaria agallocha* (N-FINDR and ATGP).

ground values in comparison to the classified result of ATGP (unconstrained and fully constrained) and PPI.

5.3.3 Accuracy Assessment

The confusion matrix shows the accuracy of classification by comparing its results with ground truth information. The confusion matrix is calculated by comparing the location and class of each ground truth pixel with the corresponding location and class in the classification image. Each column of the confusion matrix represents a ground truth class and the values in the column correspond to the classified image's labeling of the ground truth pixels.

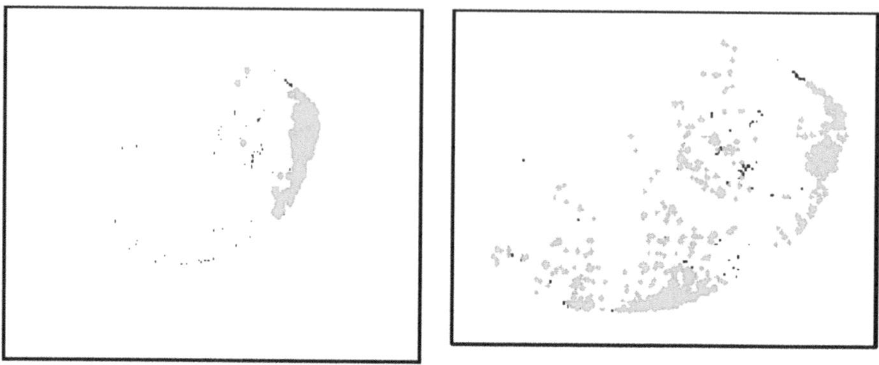

FIGURE 5.3 Unconstrained LSU fractional abundance of *Ceriops decandra* (N-FINDR and ATGP).

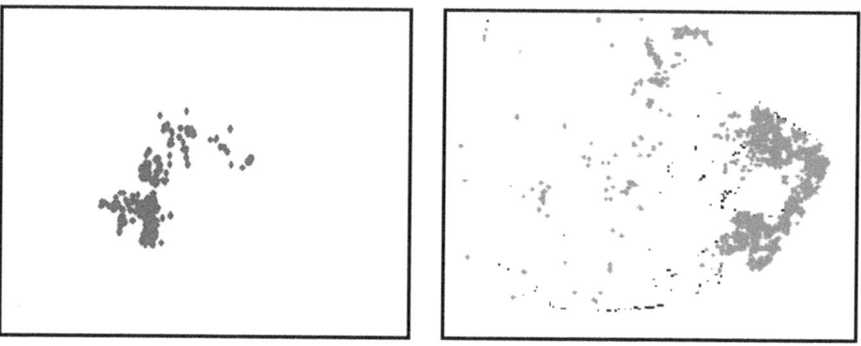

FIGURE 5.4 Unconstrained LSU fractional abundance of *Phoenix paludosa* (N-FINDR and ATGP).

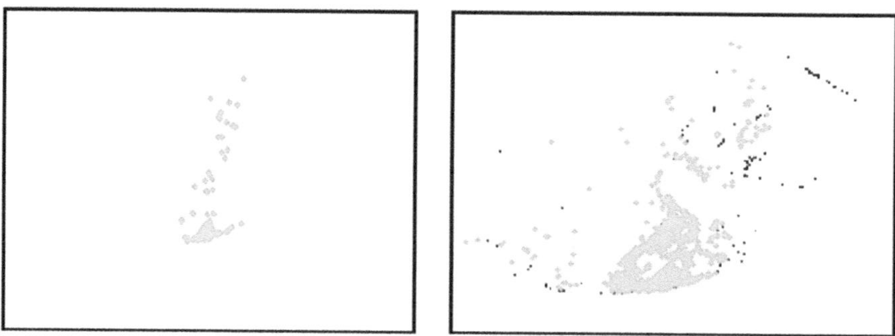

FIGURE 5.5 Unconstrained LSU fractional abundance of *Avicennia alba* (N-FINDR and ATGP).

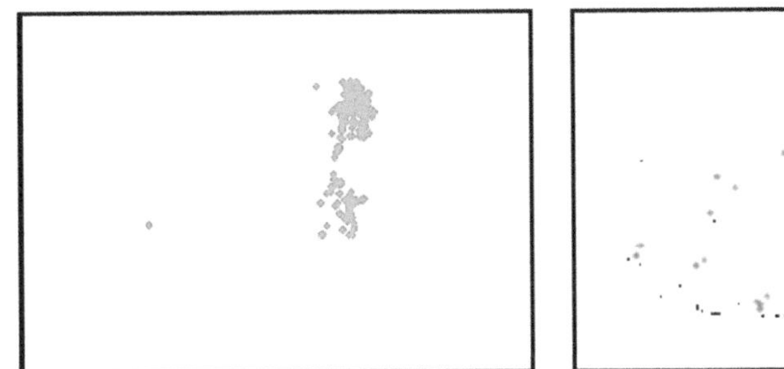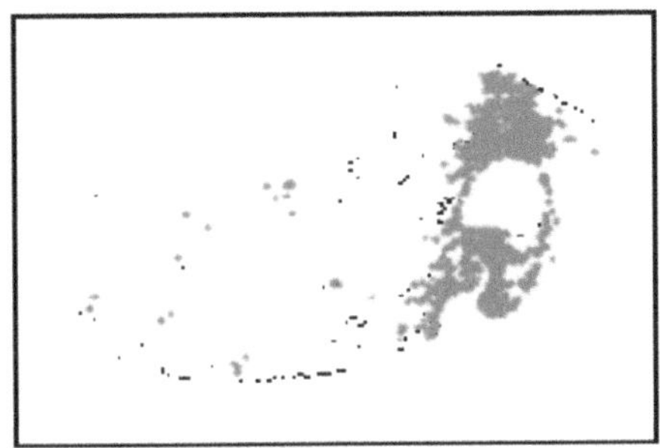

FIGURE 5.6 Unconstrained LSU fractional abundance of *Avicennia marina* (N-FINDR and ATGP).

FIGURE 5.7 Unconstrained LSU fractional abundance of *Bruguiera cylindrica* (N-FINDR and ATGP).

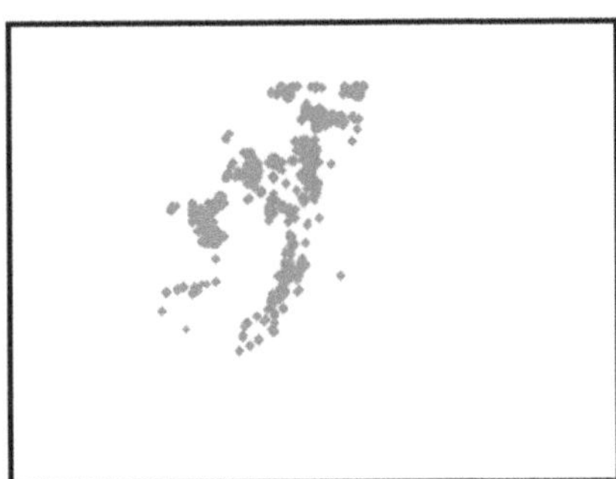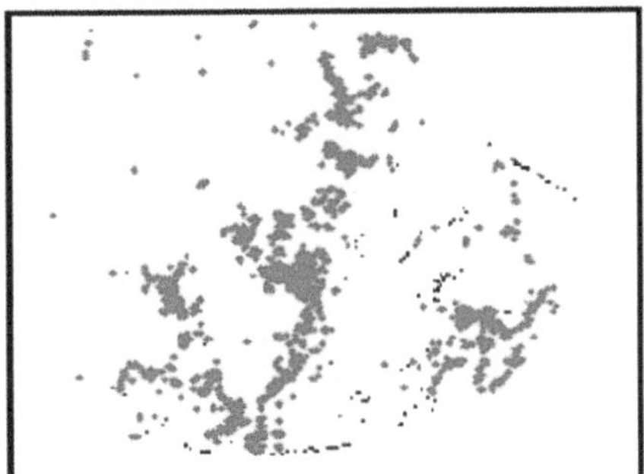

FIGURE 5.8 Unconstrained LSU fractional abundance of *Avicennia officinalis* (N-FINDR and ATGP).

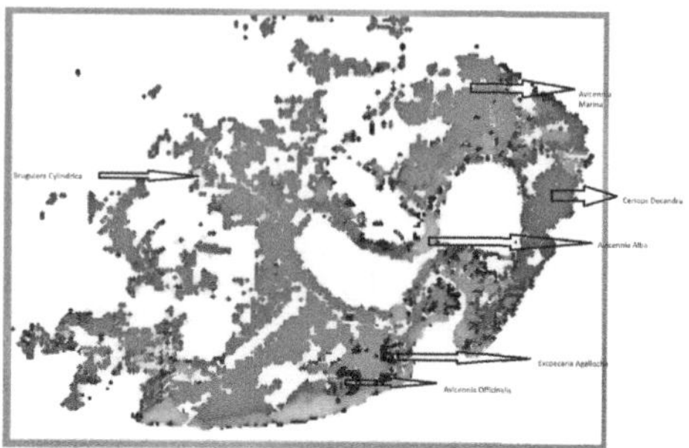

Excoecaria agallocha   *Ceriops decandra*   *Avicennia marina*

Avicennia officinalis   *Bruguiera cylindrica*   *Avicennia alba*

FIGURE 5.9   Abundance distribution of mangrove species with PPI endmembers as input.

The overall accuracy is calculated by summing the number of pixels classified correctly and dividing by the total number of pixels. The ground truth image or ground truth region of interest (ROI) defines the true class of the pixels. The pixels classified correctly are found along the diagonal of the confusion matrix table, which lists the number of pixels that are classified into the correct ground truth class (Beck, 2003). The total number of pixels is the sum of all pixels in all the ground truth classes.

The Kappa coefficient is another measure of the accuracy of classification. It is calculated by multiplying the total number of pixels in all the ground truth classes (N) by the sum of the confusion matrix diagonals (xkk), subtracting the sum of the ground truth pixels in a class times the sum of the classified pixels in that class summed over all classes(), and dividing by the total number of pixels squared minus the sum of the ground truth pixels in that class times the sum of the classified pixels in that class summed over all classes.

The accuracy of the classified output is substantiated after comparing the confusion matrix and Kappa coefficient (Tables 5.1, 5.2, 5.3, and 5.4 for N-FINDR-LSU and ATGP-LSU).

The overall accuracy of the classified output of six identified endmembers using unconstrained LSU (N-FINDR) was measured at about 74.07% with a Kappa value of 0.683. The confusion matrix of six species (*Excoecaria agallocha, Avicennia officinalis, Ceriops, Avicennia marina, Phoenix paludosa,* and *Bruguiera cylindrica*) has been shown in Table 5.1. The overall accuracy of classification from fully constrained LSU (Table 5.2) is 55.08% and the Kappa coefficient is 0.453. The strength of agreement is considered to be moderate. The accuracy of classification obtained from unconstrained and fully constrained LSU (ATGP) is very poor. The confusion matrix is shown in Tables 5.3 and 5.4, respectively. The overall accuracy achieved from unconstrained LSU (ATGP) is 23.20% with a Kappa value of 0.060. The overall accuracy achieved from unconstrained LSU (ATGP) is 22.73% with a

TABLE 5.1
Confusion Matrix of Unconstrained LSU (N-FINDR)

Class (Species Types)	Excoecaria agallocha	Avicennia officinalis	Ceriops decandra	Avicennia marina	Phoenix paludosa	Bruguiera cylindrica	Total
Excoecaria agallocha	11	1	1	1	0	0	14
Avicennia officinalis	1	32	0	0	0	0	33
Ceriops decandra	2	4	28	0	0	0	34
Avicennia marina	3	16	0	35	0	0	54
Phoenix paludosa	1	8	9	0	18	0	36
Bruguiera cylindrica	0	0	1	0	1	16	18
Total	18	61	39	36	19	16	189

TABLE 5.2
Confusion Matrix of Fully Constrained LSU (N-FINDR)

Class (Species Types)	*Excoecaria agallocha*	*Avicennia officinalis*	*Ceriops decandra*	*Avicennia marina*	*Phoenix paludosa*	*Bruguiera cylindrica*	Total
Excoecaria agallocha	7	7	0	3	0	0	17
Avicennia officinalis	4	26	0	1	0	0	31
Ceriops decandra	3	9	24	2	0	3	41
Avicennia marina	4	14	0	24	0	3	45
Phoenix paludosa	1	14	4	3	14	0	36
Bruguiera cylindrica	0	0	0	7	2	8	17
Total	19	70	28	40	16	14	187

TABLE 5.3
Confusion Matrix of Unconstrained LSU (ATGP)

Class (Species Types)	Excoecaria agallocha	Avicennia officinalis	Ceriops decandra	Avicennia marina	Phoenix paludosa	Bruguiera cylindrica	Total
Excoecaria agallocha	0	2	5	3	0	4	14
Avicennia officinalis	0	10	5	11	2	5	33
Ceriops decandra	5	7	1	7	2	5	27
Avicennia marina	0	12	7	23	4	18	64
Phoenix paludosa	6	7	5	1	11	9	39
Bruguiera cylindrica	0	0	0	0	17	0	17
Total	11	38	23	45	36	41	194

TABLE 5.4
Confusion Matrix of Fully Constrained LSU (ATGP)

Class (Species Types)	Excoecaria agallocha	Avicennia officinalis	Ceriops decandra	Avicennia marina	Phoenix paludosa	Bruguiera cylindrica	Total
Excoecaria agallocha	0	2	6	3	0	3	14
Avicennia officinalis	1	12	4	2	3	5	27
Ceriops decandra	8	2	2	11	5	0	28
Avicennia marina	0	5	6	5	0	2	18
Phoenix paludosa	6	6	0	1	11	5	29
Bruguiera cylindrica	0	0	0	0	16	0	16
Total	15	27	18	22	35	15	132

TABLE 5.5

Fractional Abundance and Root Mean Square Error Values of Pure Mangrove Patch Species

Geographical Location on Image	Ceriops Dominated Patch (21.572989 oN; 88.296710E)		
Algorithm Used	LSU with N-FINDR	LSU with PPI	Ground Truth
Excoecaria agallocha	0.3779	0.2189	0.3
Avicennia alba	0.0420	0.3502	0
Ceriops decandra	0 .4600	0.4119	0.5
Bruguiera cylindrical	0	0.0189	0
Avicennia officinalis	0.0649	0	0.1
Phoenix paludosa	0.0552	–	0.1
Avicennia marina	0	0	0
Root Mean Square Error	0.1630	0.2602	

Kappa value of 0.056. The accuracy of these algorithms is further substantiated after comparing the root mean square error values of fractional abundances for different endmembers from classification results and ground truth values (Table 5.5).

A field survey has been made at coordinates 21.572989oN; 88.296710E, which represent a dominant patch of *Ceriops decandra* (50%) with lesser abundances of *Excoecaria agallocha* (30%), *Avicennia officinalis* (10%), and *Phoenix paludosa* (10%). The fractional abundance values generated by LSU using N-FINDR and PPI as input have been displayed in Table 5.5. A comparison reveals that LSU with N-FINDR endmembers as input presently has lower RMSE values for the above-mentioned coordinate as compared with PPI members as input.

Through this study, it is observed that accurate estimation of fractional abundance of target subpixels (mangrove species in this case) is an essential part in addition to target detection of the image processing exercise. Such quantification of endmembers is required in higher-order non-linear models and sub-pixel mapping.

In conclusion, this chapter has provided a detailed review of the least-squares-based LSU method for the analysis of hyperspectral data. The mathematical formulation of the method was explained, along with its advantages and limitations. The performance of the LSU method and its application to real-world datasets was also discussed. The case study of mangrove species classification using the Hyperion dataset of Henry Island of the Sunderbans demonstrated the utility of the LSU method in fractional abundance estimation of mangrove species. The use of constrained and unconstrained LSU on the data was also discussed. Overall, this chapter highlights the potential of the LSU method in the analysis of hyperspectral data for various applications and emphasizes the importance of proper preprocessing and endmember extraction in achieving accurate results.

6 Non-Linear Unmixing for Classification of Mixed Endmembers

6.1 LIMITATIONS OF LINEAR SPECTRAL UNMIXING

Linear spectral unmixing (LSU) methods, such as the least-squares-based LSU method, have been widely used for the analysis of hyperspectral data. However, these methods have several limitations that can affect their accuracy and applicability. One major limitation is the assumption of pure endmembers, which is often not met in real-world scenarios. In many cases, the endmembers in the scene are not pure, but rather are mixtures of other endmembers, leading to the presence of mixed pixels. These mixed pixels cannot be accurately represented by a linear combination of the pure endmembers, resulting in an underestimation of the fractional abundances of the endmembers.

Another limitation of LSU methods is the assumption of linearity in the mixing process. In many cases, the mixing process is non-linear, leading to a non-unique solution in the unmixing process. This can lead to errors in the estimation of the fractional abundances and can also lead to multiple solutions with different fractional abundances.

In addition, LSU methods also assume that the spectra of the endmembers do not overlap, which is also not always the case. Spectral overlap can lead to ambiguities in the unmixing process, resulting in errors in the fractional abundances.

To overcome these limitations, NLU methods have been developed. These methods are based on more complex mathematical models that take into account the non-linearity and overlap in the mixing process. Examples of NLU methods include Nascimento's fully constrained least squares (FCLS), Fan's, Hapke, and higher-order models.

6.2 NON-LINEAR UNMIXING MODELS

Non-linear unmixing models are advanced techniques for analyzing hyperspectral data that are designed to overcome the limitations of linear unmixing methods. These models take into account the complex interactions between different types of materials and their spectral signatures and can provide more accurate results in cases where linear methods fail.

DOI: 10.1201/9781003432623-6

Non-linear models are applied for entities that are intricately mixed (Heylen, 2014; Hapke, 1993). Non-linear interactions result from scattering of light by a given surface, which subsequently gets reflected from other surfaces before reaching the sensor. Non-linear interactions occurring at macroscopic scale have been studied in multilayered configurations (Halimi et al., 2011; Chen et al., 2013). This is often the case for forested areas, where there are many interactions between the ground surface and the canopy

6.3 NASCIMENTO'S BILINEAR SPECTRAL UNMIXING MODEL

Nascimento's bilinear spectral unmixing model is a non-linear unmixing method that attempts to overcome the limitations of linear unmixing methods. The model is based on the assumption that the spectral signature of each pixel in a hyperspectral image is a linear combination of the endmember spectra but with non-linear mixing coefficients. In the bilinear model developed by Nascimento and Bioucas-Dias (2009; 2010; 2012), it is assumed that the light falling on each endmember interacts with other species present in a close neighborhood.

This interaction results in a change in the reflectance spectrum of the endmember, which is modeled as a bilinear combination of the pure endmember spectrum and a mixing matrix. The mixing matrix represents the contribution of the neighboring endmembers to the observed spectrum. The bilinear model can be represented mathematically as

$$y = Mx + E$$

where y is the observed spectrum, M is the mixing matrix, x is the fractional abundance vector, and E is the error term. The bilinear model is solved using an alternating least-squares (ALS) algorithm. The ALS algorithm alternately estimates the fractional abundance vector x and the mixing matrix M, until convergence is achieved.

One of the advantages of the bilinear model is that it can handle mixed pixels, where multiple endmembers are present in different proportions. It also allows for the estimation of the noise term E, which can be used to identify pixels with high noise levels. The bilinear model has been applied to various hyperspectral datasets, including vegetation, minerals, and urban environments, and has shown good performance in unmixing mixed pixels and estimating noise levels.

An example of this interaction could be when light reflects off of a tree canopy and also interacts with the leaves and branches of surrounding trees. This interaction causes a mixing of the reflectance spectra, resulting in a mixed endmember. Another example could be when light reflects off of a body of water and also interacts with floating algae, resulting in a mixed endmember. This model takes into account these interactions by considering the mixing of endmembers as a bilinear process rather than a linear one.

6.4   FAN'S BILINEAR UNMIXING MODEL

The bilinear unmixing model developed by Fan and his colleagues (Fan et al., 2009) is a generalization of the linear unmixing model, where the abundance fractions of each endmember are modeled as a function of the endmember's reflectance and the reflectance of other endmembers. The model is formulated as

$$Y = A * X + E$$

where Y is the measured reflectance, X is the matrix of endmember reflectances, A is the matrix of endmember abundance fractions, and E is the measurement noise. The model is non-linear because the abundance fractions in A are modeled as a function of X.

The Fan bilinear unmixing model solves this problem by assuming that the abundance fractions are a function of both the endmember reflectances and the reflectances of other endmembers. Specifically, the model assumes that the abundance fractions of each endmember are a bilinear function of the endmember reflectances and the reflectances of other endmembers.

The model is formulated as

$$Y = A * X + E$$

where Y is the measured reflectance, X is the matrix of endmember reflectances, A is the matrix of endmember abundance fractions, and E is the measurement noise. The model is non-linear because the abundance fractions in A are modeled as a function of X.

The Fan bilinear unmixing model can be solved using a variety of optimization algorithms, such as gradient descent or Newton's method. In addition, the model can be regularized using various techniques, such as non-negativity constraints, sparsity constraints, or smoothness constraints, to improve the stability and interpretability of the solution.

One of the advantages of the Fan bilinear unmixing model is that it can handle endmembers that are correlated, meaning that their reflectance spectra are highly similar. This is because the model takes into account the reflectance of other endmembers when estimating the abundance fractions of each endmember, which allows it to better handle correlated endmembers than linear unmixing methods.

An example of the application of the Fan bilinear unmixing model is in the classification of mixed pixels in hyperspectral data. In this application, the model can be used to estimate the abundance fractions of each endmember in mixed pixels, which can then be used to classify the mixed pixels into different classes based on their endmember composition. This can be a more accurate and robust way to classify mixed pixels than linear unmixing methods, especially when the endmembers are correlated.

6.5   HAPKE'S BIDIRECTIONAL MODEL

The Hapke model, developed by Bruce Hapke in the early 1980s, is a non-linear unmixing model that describes the bidirectional reflectance of a surface as a function of the incidence angle, azimuth angle, and phase angle of the light. The model takes into account the effects of surface roughness, shadowing, and subsurface scattering on the reflectance of the surface.

The following equation gives the Hapke model:

$$r(\theta, \phi, \gamma) = (B_0 + B_1 w) + (H_0 + H_1 w)\cos(\theta) + (G_0 + G_1 w)\cos^2(\theta)$$

where

$r(\theta, \phi, \gamma)$ is the bidirectional reflectance of the surface as a function of the incidence angle (θ), azimuth angle (ϕ), and phase angle (γ).

B_0 and B_1 are the isotropic and anisotropic reflectance components, respectively.

H_0 and H_1 are the isotropic and anisotropic components of the opposition effect, respectively.

G_0 and G_1 are the isotropic and anisotropic components of the shadowing effect, respectively.

w is the single-scattering albedo.

The Hapke model has several adjustable parameters, including the opposition effect parameter (B_0), the anisotropy parameter (B_1), the shadowing parameter (G_0), and the single-scattering albedo (w). These parameters can be estimated using a variety of optimization techniques, such as least-squares or maximum likelihood estimation.

One of the advantages of the Hapke model is that it can be used to estimate the fractional abundances of the materials in a scene, even when the materials have similar spectral signatures. This is because the model takes into account how the materials scatter light, which can be a unique characteristic of the material.

An example of the Hapke model being applied to a dataset is in analyzing the mineral composition of the moon's surface. The Clementine spacecraft collected hyperspectral data of the lunar surface, which was then analyzed using the Hapke model to estimate the fractional abundances of the minerals present.

Overall, the Hapke model is a powerful tool for the non-linear unmixing of mixed endmembers, and it can be used to analyze a wide range of datasets, including mineral mapping, vegetation analysis, and soil analysis. However, it does require accurate laboratory measurements or field observations to determine the empirical parameters. It can be computationally intensive and requires a good understanding of the physics of light scattering.

6.6 HIGHER-ORDER NON-LINEAR SPECTRAL UNMIXING MODELS

Higher-order non-linear models are a class of models that go beyond the traditional linear and bilinear models by incorporating higher-order interactions between the endmembers and the mixed pixels. These models are based on polynomial functions and can capture more complex interactions between the endmembers in a mixed pixel.

One example of a higher-order non-linear model is the trilinear model, which incorporates a cubic term in the model equation. This allows for the inclusion of interactions between three endmembers in a mixed pixel rather than just two, as in the bilinear model.

Another example is the quadrilinear model, which includes a quartic term in the model equation. This allows for the inclusion of interactions between four endmembers in a mixed pixel.

A key advantage of higher-order non-linear models is their ability to capture more complex interactions between endmembers, which can result in a more accurate estimation of the fractional abundances of the endmembers in a mixed pixel. However, a disadvantage is that these models can be computationally intensive and may require a more significant number of endmembers to be accurately estimated.

One example of the application of higher-order models is the study of vegetation by using Hyperion data. In this study, a quadrilinear model was applied to the Hyperion data to estimate the fractional abundances of four vegetation endmembers (evergreen needleleaf forest, evergreen broadleaf forest, deciduous needleleaf forest, and deciduous broadleaf forest) in a mixed pixel. The results showed that the quadrilinear model outperformed linear and bilinear models in terms of accuracy and precision, demonstrating the benefit of higher-order models in more complex scenarios.

Higher-order non-linear models are extensions of the bilinear models, where the assumptions are made that the endmembers are not pure and they interact with each other in more complex ways. These models typically involve more than two terms in the mixing matrix, and polynomial functions of the reflectance values can represent them.

One example of a higher-order non-linear model is the fourth-order polynomial model proposed by Chang et al. (2015). This model assumes that the reflectance of each pixel is a fourth-order polynomial function of the endmember reflectances, and it can be written as

$$
\begin{aligned}
R(\lambda) = {}& a1E1(\lambda) + a2E2(\lambda) + a3E3(\lambda) + a4E4(\lambda) + a5E1(\lambda)E2(\lambda) + a6E1(\lambda)E3(\lambda) \\
& + a7E1(\lambda)E4(\lambda) + a8E2(\lambda)E3(\lambda) + a9E2(\lambda)E4(\lambda) + a10E3(\lambda)E4(\lambda) \\
& + a11E1(\lambda)E2(\lambda)E3(\lambda) + a12E1(\lambda)E2(\lambda)E4(\lambda) + a13E1(\lambda)E3(\lambda)E4(\lambda) \\
& + a14E2(\lambda)E3(\lambda)E4(\lambda) + a15E1(\lambda)E2(\lambda)E3(\lambda)E4(\lambda)
\end{aligned}
$$

where $R(\lambda)$ is the reflectance of the mixed pixel, $Ei(\lambda)$ is the reflectance of the ith endmember, and ai are the mixing coefficients.

Another example of a higher-order non-linear model is the Tucker3 model, which is based on a multiway array decomposition approach, and it was proposed by Tucker (1966). This model can be represented by a three-way array, where the first mode represents the endmembers, the second mode represents the abundance maps, and the third mode represents the interaction between endmembers and abundance maps. The Tucker3 model can be written as

$$
R = G \times A \times E'
$$

where R is the multiway array of the mixed pixels, G is a matrix of interaction coefficients, A is a matrix of abundance maps, E' is a matrix of endmembers, and $\times$ represents the mode-n product.

Itis important to note that these models are complex and computationally intensive. They also require a large amount of data and accurate initialization of parameters.

In conclusion, higher-order non-linear models are a class of models that go beyond the traditional linear and bilinear models by incorporating higher-order interactions between the endmembers and the mixed pixels. These models are based on polynomial functions and can capture more complex interactions between the endmembers in a mixed pixel, resulting in a more accurate estimation of the fractional abundances of the endmembers in a mixed pixel. However, these models can be computationally intensive and may require a more significant number of endmembers to be accurately estimated.

6.7 EXPERIMENTAL RESULTS OF ANALYSIS OF LOWER TO HIGHER-ORDER NON-LINEAR MODELS WITH CASE STUDIES

In this section, we have presented experimental results of applying different non-linear unmixing models, including Nascimento's bilinear model and Fan's bilinear model to real-world hyperspectral datasets. We have also included results for higher-order models, such as the quadratic and cubic models. The datasets used have been discussed in detail, including the acquisition parameters and the ground truth information. The results have been presented in abundance maps, classification maps, and quantitative measures such as overall accuracy and Kappa coefficient. We have also provided a comparison of the different models' performance, highlighting each advantages and limitations.

The N-FINDR extracted endmembers have been taken as input for the execution of non-linear models, namely Nascimento's bilinear model, higher-order models for intra-species interaction upto second order, higher-order models for inter-species interaction up to third and fourth orders (Figures 6.2–6.21). In this way, a total of seven dominant mangrove species have been identified in the study area (Figure 6.1).

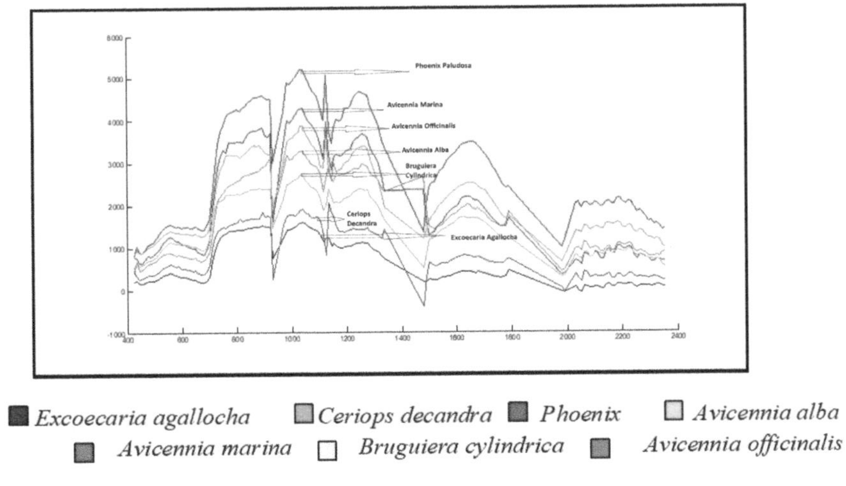

FIGURE 6.1 Spectral profile of mangrove species identified with the N-FINDR algorithm.

It is then assumed that these mixed mangrove stands are admixtures of one/more of the identified species types.

Nascimento's bilinear model of second order is at first executed, which yielded 28 interactions (seven single or linear interactions and 21 inter-species interactions).

The higher-order model for studying intra-species interactions up to second order is next executed, which resulted in finding 35 interactions (seven single or linear interactions, 21 bilinear interactions, and seven intra-species interactions).

The higher-order model for inter-species interactions of third order has identified 70 interactions (seven linear interactions, 21 bilinear interactions, seven intra-species interactions, and 35 inter-species interactions of third order). It may be mentioned here that the higher-order model of second, third, fourth, and Nth order are represented by equations that have additive functions to indicate linear and intra-species interactions. An occurrence of a pure mangrove patch reduces the equations to the linear and intra-species interaction part of the model.

Tables 6.1 and 6.2 show the fractional abundance estimates and the RMSE values of a mixed patch of *Excoecaria agallocha-Avicennia officinalis-Bruguiera cylindrica-Avicennia alba*. It is observed that the fourth-order model, and then the third-order model, is most accurate for the classification of mixed stands of 4-3 mangrove species, respectively. As the linear spectral unmixing model does not consider mixture interactions at all, it shows the least accuracy. Nascimento's bilinear model considers only two endmember mixture interactions; hence it shows lower accuracy as compared to third- and fourth-order models. Abundance calculation shows the negligible presence of mixed stands with four species (endmembers) together.

A comparison of fractional abundance images generated with the five models, linear spectral unmixing, Nascimento's bilinear model, higher-order model for intra-species interaction upto second order, higher-order model for inter-species interaction (up to third and fourth order) for *Excoecaria agallocha-Avicennia officinalis-Bruguiera cylindrica-Avicennia alba* mixed patch and their interactions for the above coordinate is shown in Figures 6.2–6.21. The figures display pure pixels of mangrove species having a fractional abundance of 50% and above. Figures 6.2, 6.3, and 6.4 display abundance of *Excoecaria*pure patch with LSU, abundance of *Excoecaria*pure patch with Nascimento's bilinear model and abundance of *Excoecaria*pure patch with HONLUM for intra-species interaction of second Order. Figures 6.5, 6.6, and 6.7 display abundance of *Avicennia alba*pure patch with LSU, abundance of *Avicennia alba*pure patch with Nascimento's bilinear model and abundance of *Excoecaria*pure patch with HONLUM for intra-species interaction of second Order. Figures 6.8, 6.9, and 6.10 display abundance of *Bruguiera cylindrica*pure patch with LSU, abundance of *Bruguiera cylindrica*pure patch with Nascimento's bilinear model and abundance of *Bruguiera cylindrica*pure patch with HONLUM for intra-species interaction of secondorder. Figures 6.11, 6.12, and 6.13 display abundance of *Avicennia officinalis*pure patch with LSU, abundance of *Avicennia officinalis*pure patch with Nascimento's bilinear model and abundance of *Avicennia officinalis*pure patch with HONLUM for intra-species

TABLE 6.1

Comparative Analysis of Different Model Outputs in Relation to Mixed Mangrove Patches: Data Relates to Fractional Abundance Values and Root Mean Square Error at Geographic Coordinates 21°34.621′N; 88°16.502′E

	Fractional Abundance Values					
Endmembers	Linear Spectral Unmixing Model	Nascimento's Bilinear Model	Higher Order Model of Second Order	Higher Order Model of Third Order	Higher Order Model of Fourth Order	Ground Truth Results
1 (*Excoecaria agallocha*)	0.5833	0.3710	0.2412	0.2274	0.2264	0.24
4 (*Avicennia alba*)	0.1009	0.0587	0.0369	0.0406	0.0404	0.04
6 (*Bruguiera cylindrica*)	0.1110	0.0657	0.0499	0.0469	0.0467	0.05
7 (*Avicennia officinalis*)	0.2048	0.1299	0.0810	0.0804	0.0800	0.08
1,1	–	–	0.2916	0.2604	0.2592	0.24
1,4	–	0.0713	0.0446	0.0465	0.0463	0.05
1,6	–	0.0799	0.0603	0.0537	0.0534	0.05
1,7	–	0.1579	0.0979	0.0920	0.0916	0.09
4,4	–	–	0.0068	0.0083	0.0083	0.01
4,6	–	0.0126	0.0092	0.0096	0.0095	0.01
4,7	–	0.0250	0.0150	0.0164	0.0164	0.02
6,6	–	–	0.0125	0.0111	0.0110	0.02
6,7	–	0.0280	0.0202	0.0190	0.0189	0.02
7,7	–	–	0.0329	0.0325	0.0324	0.03
1,4,6	–	–	–	0.0110	0.0109	0.01
1,4,7	–	–	–	0.0188	0.0187	0.02
1,6,7	–	–	–	0.0217	0.0216	0.03
4,6,7	–	–	–	0.0039	0.0039	0.01
1,4,6,7				–	0.0044	0.005
Root Mean Square Error	0.2315	0.0761	0.0398	0.0211	0.0188	

TABLE 6.2

Comparative Analysis of Different Model Outputs in Relation to Mixed Mangrove Patches: Data Relates to Fractional Abundance Values and Root Mean Square Error at Geographic Coordinates 21o34.374′N; 88o16.810′E

| | Fractional Abundance Values | | | | | |
Endmembers	Linear Spectral Unmixing Model	Nascimento's Bilinear Model	Higher Order Model of Second Order	Higher Order Model of Third Order	Higher Order Model of Fourth Order	Ground Truth Results
1 (*Excoecaria agallocha*)	0.2164	0.1368	0.1069	0.0928	0.0921	0.10`
2 (*Ceriops decandra*)	0.2954	0.1747	0.1363	0.1290	0.1279	0.13
3 (*Phoenix paludosa*)	0.1900	0.1128	0.0891	0.0829	0.0822	0.08
5 (*Avicennia marina*)	0.2982	0.1777	0.1395	0.1304	0.1292	0.13
1,1			0.0431	0.0349	0.0347	0.03
1,2		0.0708	0.0550	0.0486	0.0482	0.05
1,3		0.0457	0.0359	0.0312	0.0310	0.04
1,5		0.0720	0.0563	0.0491	0.0487	0.05
2,2			0.0701	0.0675	0.0669	0.07
2,3		0.0583	0.0458	0.0434	0.0430	0.04
2,5		0.0919	0.0717	0.0682	0.0676	0.07
3,3			0.0299	0.0279	0.0277	0.03
3,5		0.0593	0.0469	0.0438	0.0435	0.04
5,5			0.0734	0.0689	0.0683	0.07
1,2,3				0.0163	0.0162	0.01
1,2,5				0.0257	0.0254	0.02
1,3,5				0.0165	0.0164	0.01
2,3,5				0.0229	0.0227	0.02
1,2,3,5					0.0086	0.01
Root Mean Square Error	0.2018	0.1034	0.0266	0.0161	0.0147	

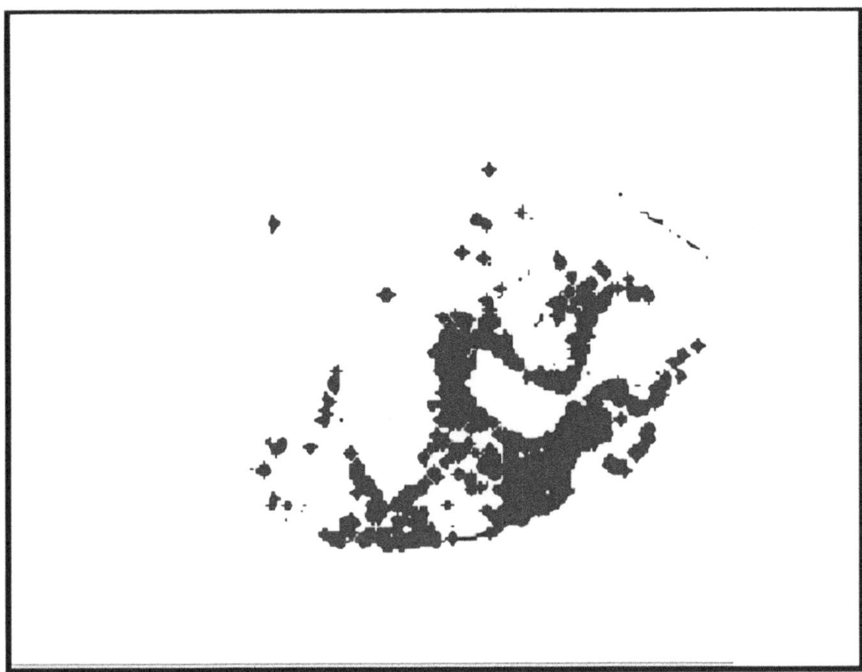

FIGURE 6.2 Abundance of *Excoecaria* pure patch with LSU.

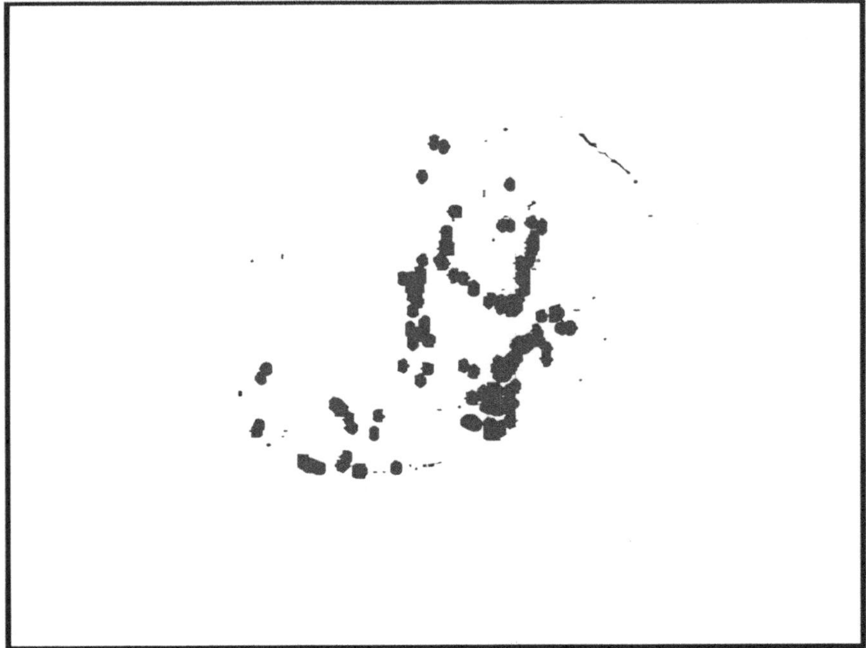

FIGURE 6.3 Abundance of *Excoecaria* pure patch with Nascimento's bilinear model.

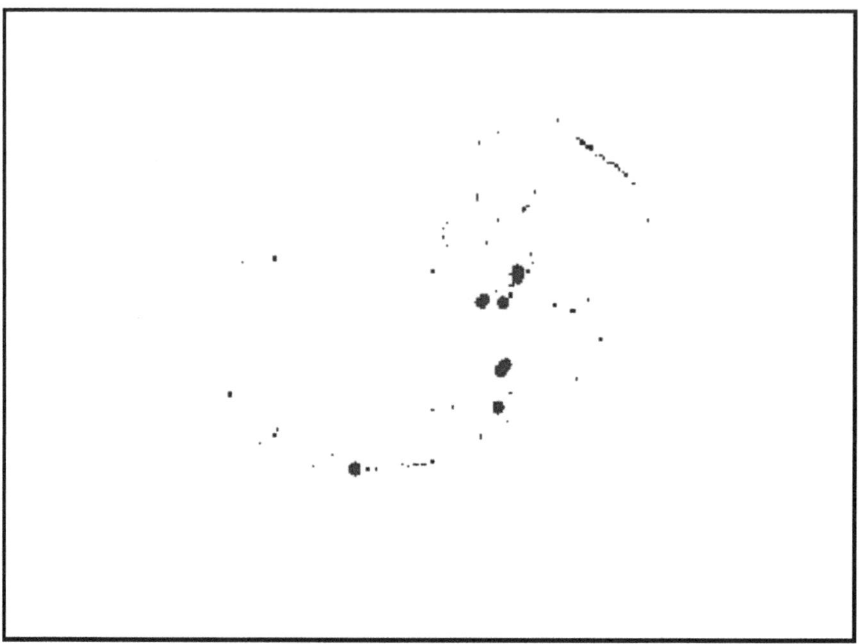

FIGURE 6.4 Abundance of *Excoecaria* pure patch with higher-order model for intra-species interaction of second order.

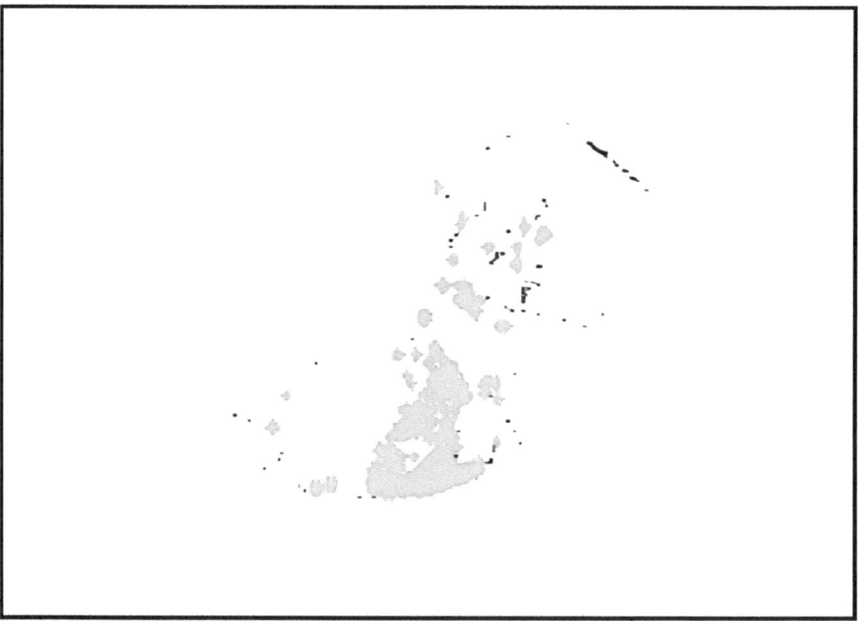

FIGURE 6.5 Abundance of *Avicennia alba* pure patch with LSU.

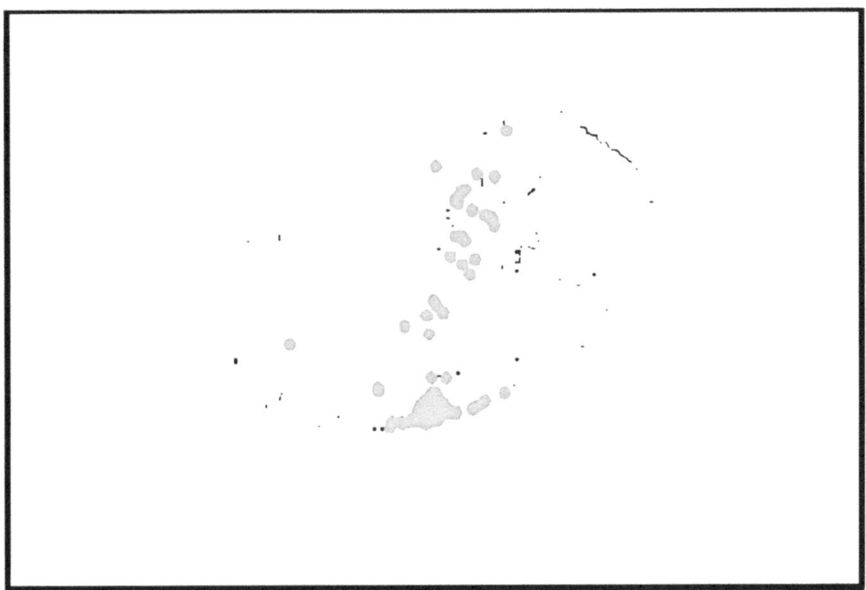

FIGURE 6.6 Abundance of *Avicennia alba* pure patch with Nascimento's bilinear model.

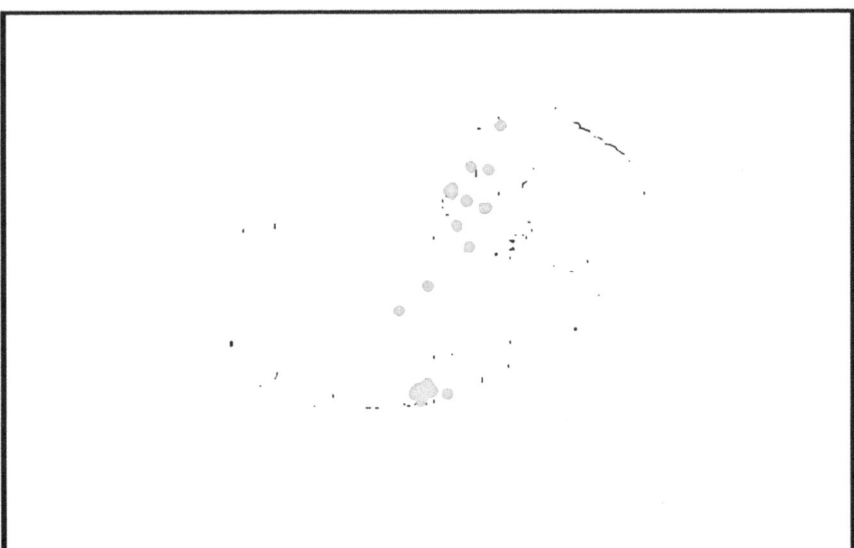

FIGURE 6.7 Abundance of *Avicennia alba* pure patch with higher-order model for intra-species interaction of second order.

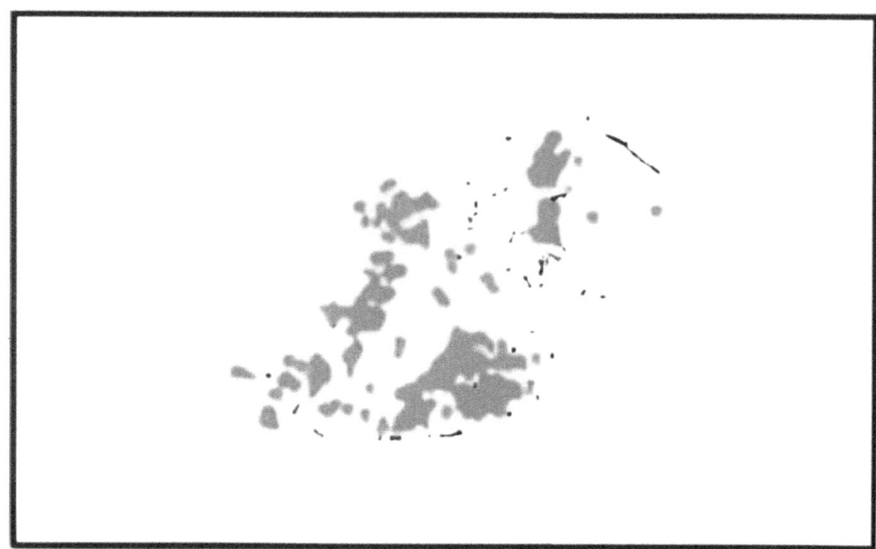

FIGURE 6.8 Abundance of *Bruguiera cylindrica* pure patch with LSU.

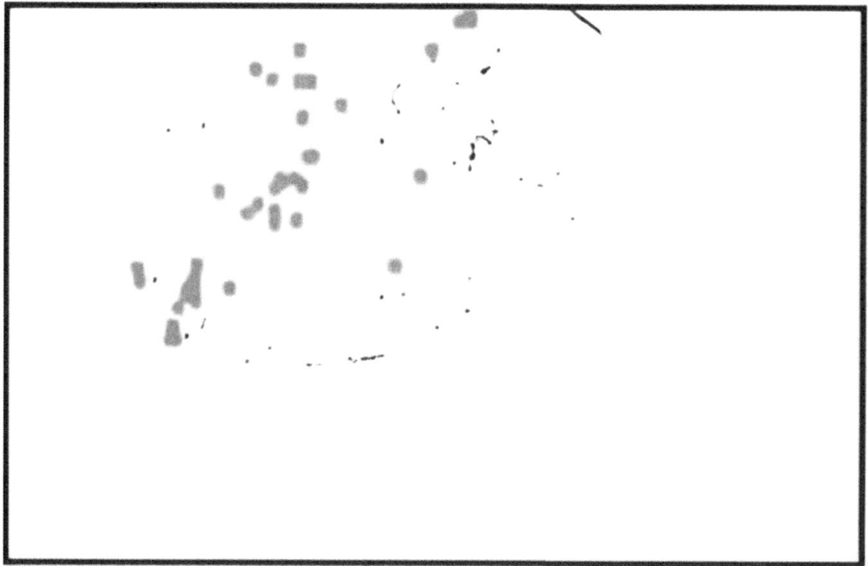

FIGURE 6.9 Abundance of *Bruguiera cylindrica* pure patch with Nascimento's bilinear model.

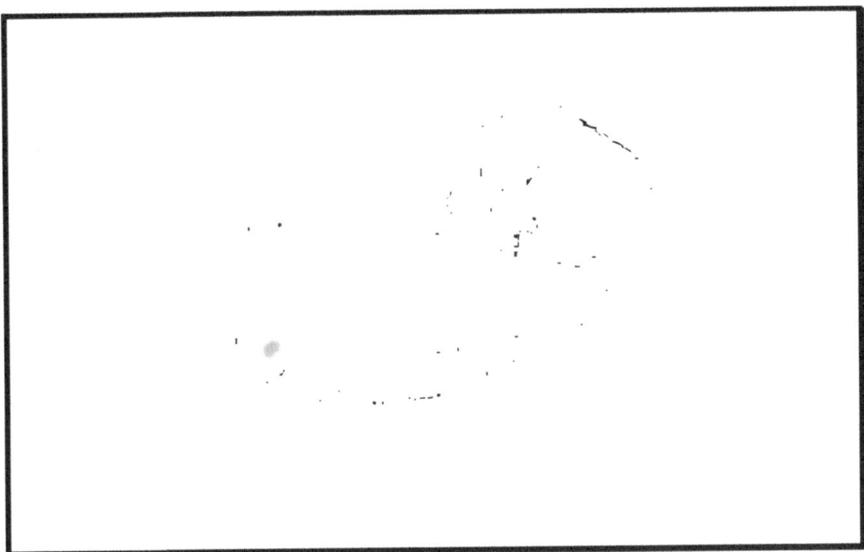

FIGURE 6.10 Abundance of *Bruguiera cylindrica* pure patch with higher-order model for intra-species interaction of second order.

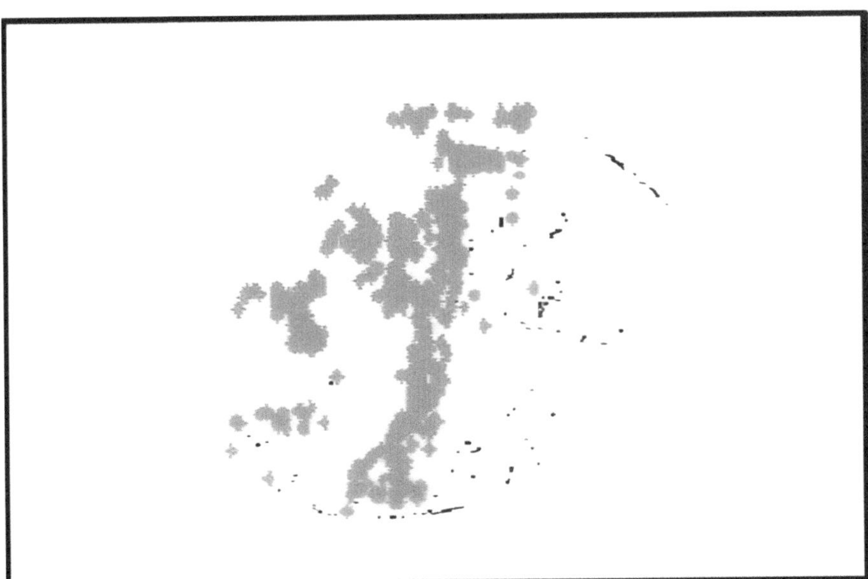

FIGURE 6.11 Abundance of *Avicennia officinalis* pure patch with LSU.

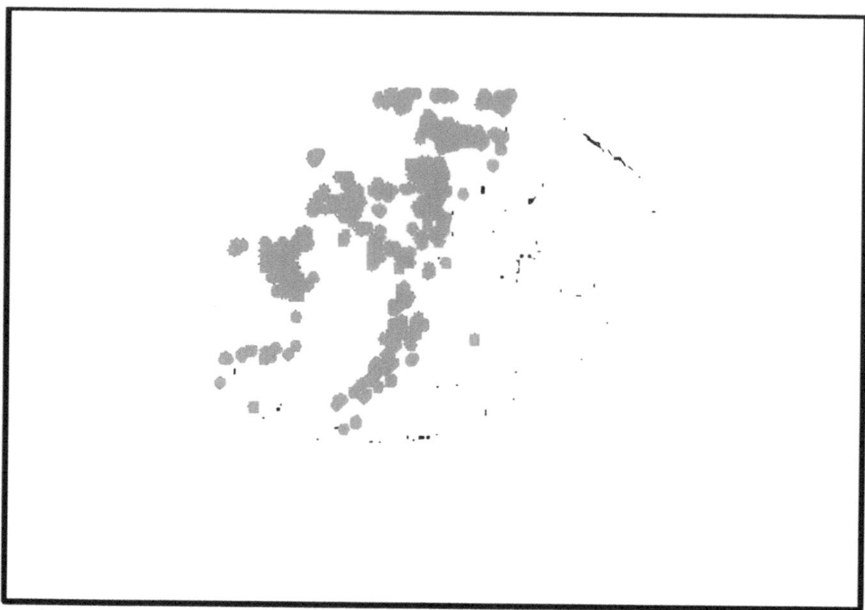

FIGURE 6.12 Abundance of *Avicennia officinalis* pure patch with Nascimento's bilinear model.

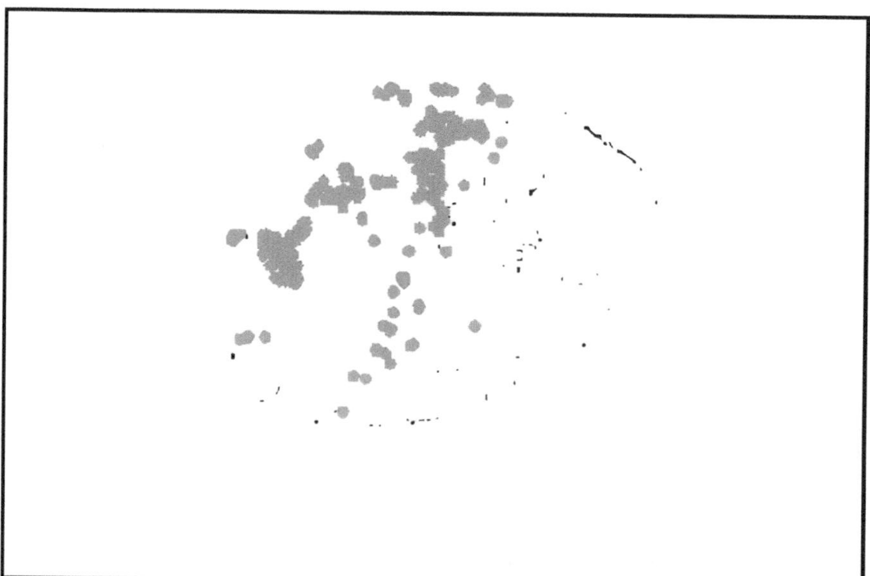

FIGURE 6.13 Abundance of *Avicennia officinalis* pure patch with higher-order model for intra-species interaction of second order.

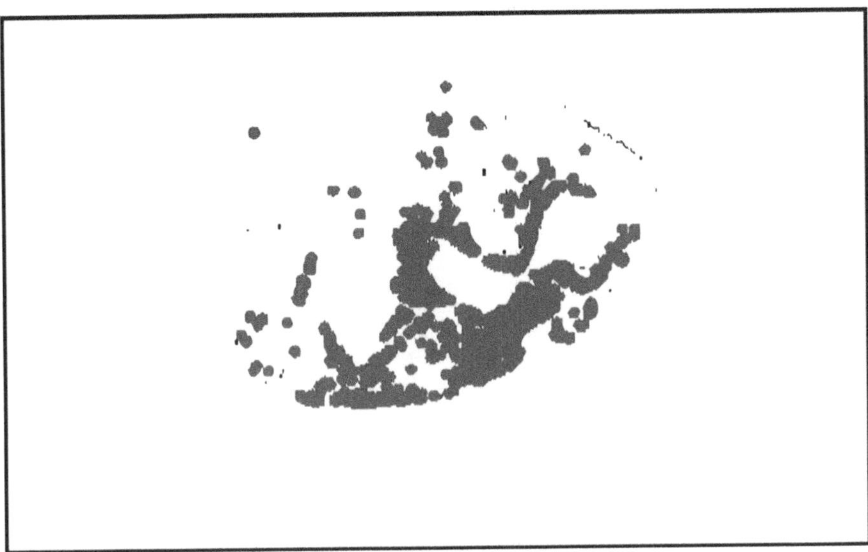

FIGURE 6.14 Abundance (8% and above) of *Excoecaria agallocha* for higher-order model for intra-species interaction of second order.

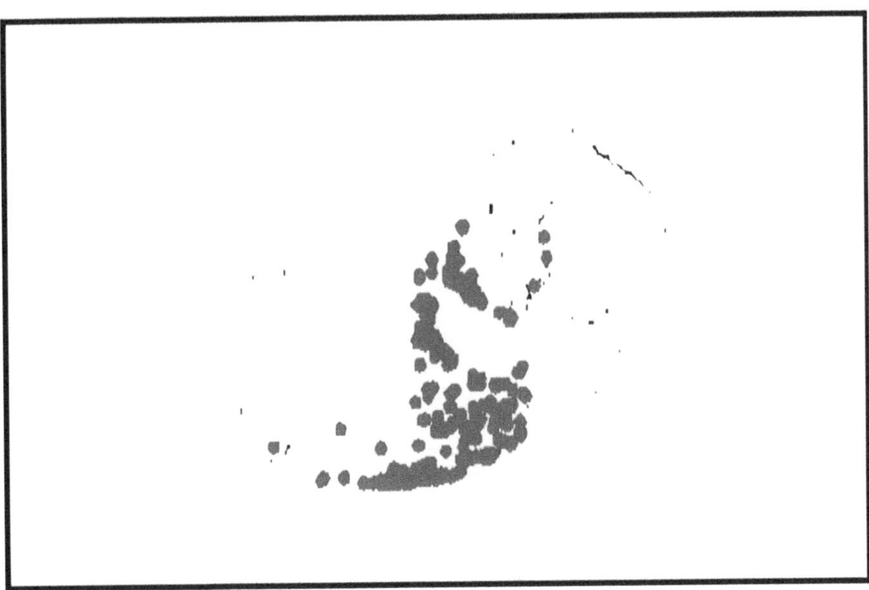

FIGURE 6.15 Abundance (1% and above) of *Avicennia alba* of higher-order model for intra-species interaction of second order.

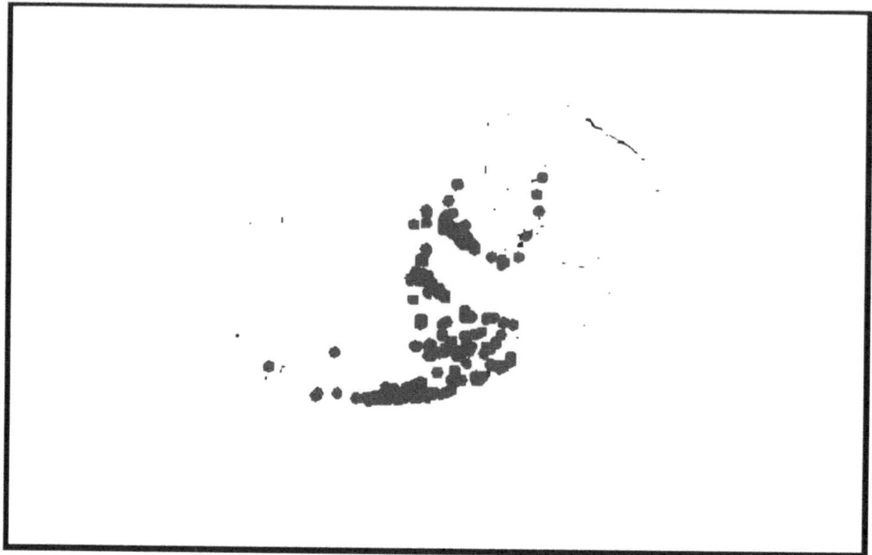

FIGURE 6.16 Abundance (8% and above) of *Excoecaria agallocha* and *Avicennia alba* mixture higher-order model for intra-species interaction of second order.

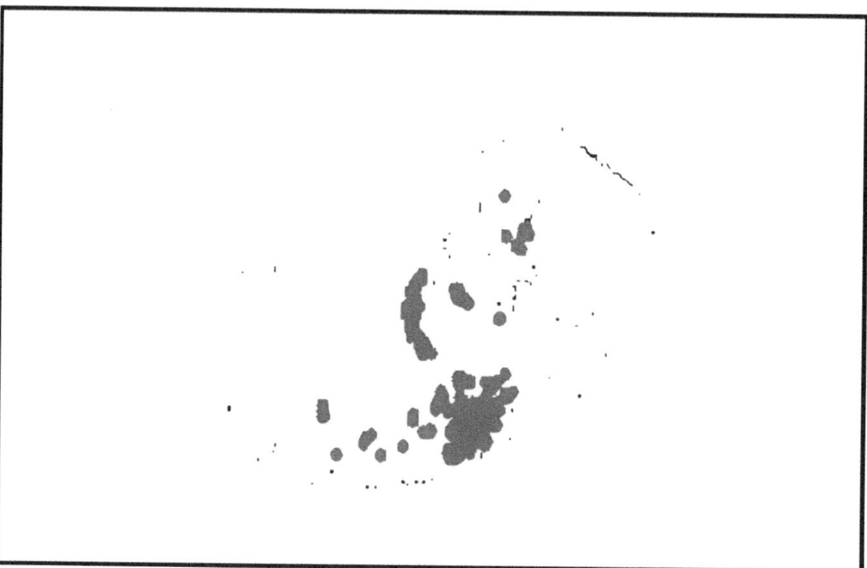

FIGURE 6.17 Abundance (1% and above) of *Excoecaria agallocha* and *Bruguiera cylindrica* mixture with Nascimento's bilinear model.

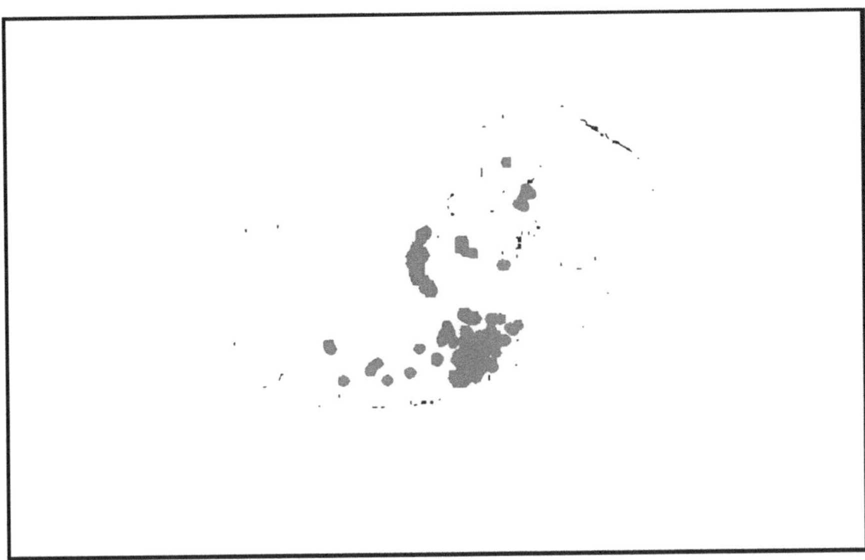

FIGURE 6.18 Abundance (8% and above) of *Excoecaria agallocha* and *Bruguiera cylindrica* mixture with higher-order model for intra-species interaction of second order.

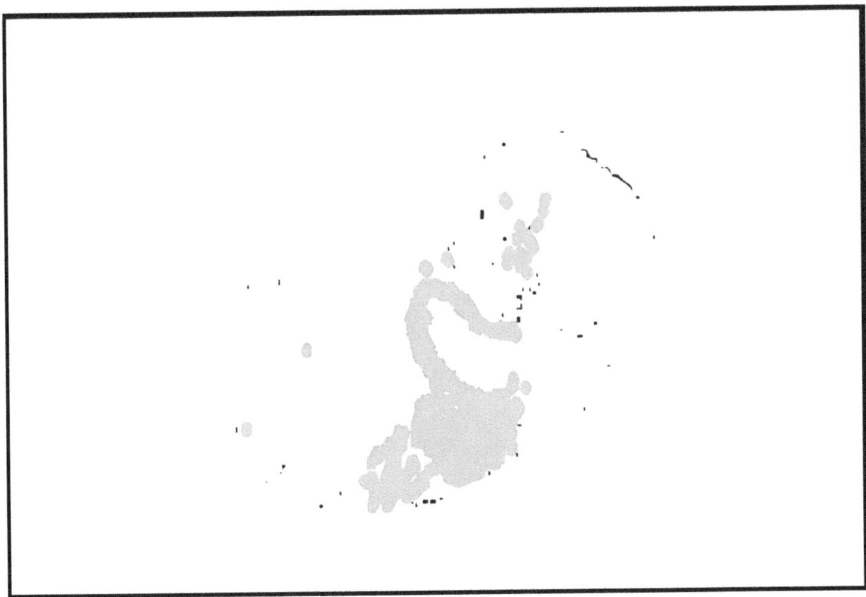

FIGURE 6.19 Abundance (1% and above) of *Excoecaria agallocha*, *Avicennia alba* and *Bruguiera cylindrica* mixture with higher-order model for inter-species interaction of third order.

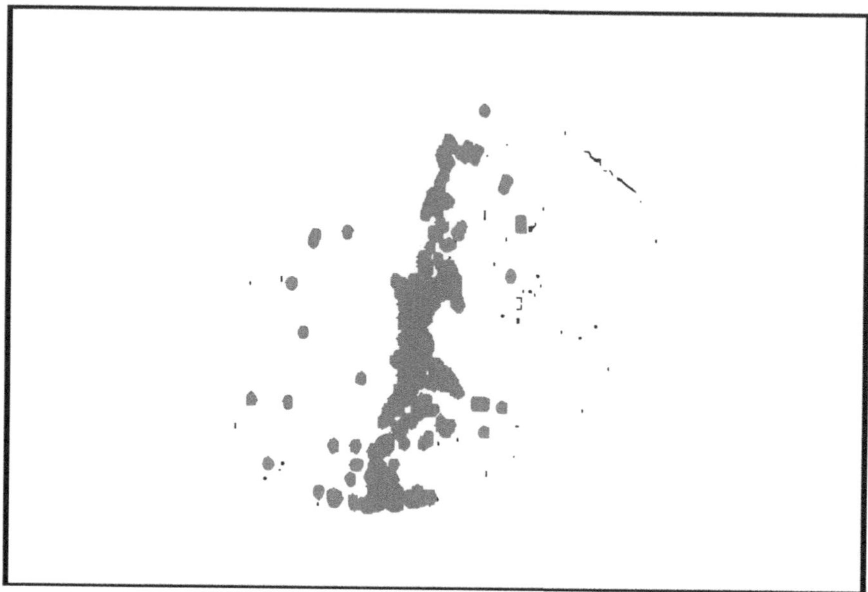

FIGURE 6.20   Abundance (1% and above) of *Excoecaria, Bruguiera cylindrica* and *officinalis* mixture with higher-order model of inter-species interaction of third order.

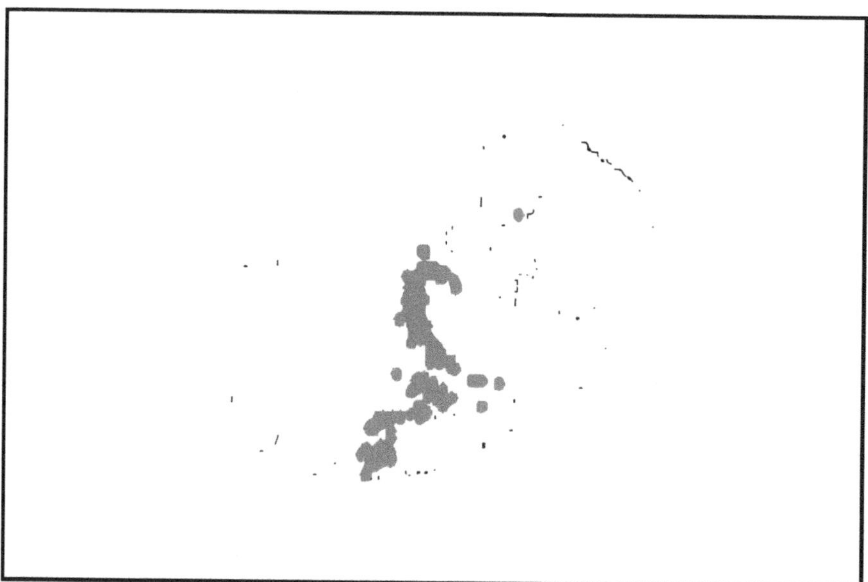

FIGURE 6.21   Abundance (0.3% and above) of *Excoecaria, alba, Bruguiera cylindrica* and *officinalis* mixture with higher-order model of inter-species interaction of fourth order.

interaction of secondorder. Figure 6.14 displays the abundance (8% and above) of *Excoecaria agallocha* for HONLUM for intra-species interaction of secondorder. Figure 6.15 displays the abundance (1% and above) of *Avicennia alba* of HONLUM for intra-species interaction of secondorder. Figure 6.16 displays the abundance (8% and above) of *Excoecaria agallocha* and *Avicenniaalba* mixture HONLUM for intra-species interaction of secondorder. Figure 6.17 displays the abundance (1% and above) of *Excoecariaagallocha* and *Bruguiera cylindrica* mixture with Nascimento's bilinear model. Figure 6.18 displays the abundance (8% and above) of *Excoecariaagallocha* and *Bruguiera cylindrica* mixture with HONLUM for intra-species interaction of secondorder. Figure 6.19 displays the abundance (1% and above) of *Excoecaria agallocha*, *Avicennia alba* and *Bruguiera cylindrica* mixture with HONLUM for inter-species interaction of third order. Figure 6.20 displays the abundance (1% and above) of *Excoecaria*, *Bruguiera cylindrica*, and *officinalis* mixture with HONLUM of inter-species interaction of thirdorder. Figure 6.21 displays the abundance (0.3% and above) of *Excoecaria*, *alba*, *Bruguiera cylindrica* and *officinalis* mixture with HONLUM of inter-species interaction of fourthorder.

Table 6.2 displays the fractional abundance estimates of *Excoecaria agallocha-Ceriops decandra-Phoenix paludosa-Avicennia marina* mixed pixel patch for the geographic coordinate 21°34.374', 88°16.810'. Table 6.3 displays the abundance estimates and RMSE values of *Excoecaria agallocha-Avicennia officinalis-Bruguiera cylindrica-Avicennia alba* mixed patch for geographic coordinate 21°34.616', 88°16.512'.

Finally, Table 6.4 shows the abundance estimate and RMSE value of *Excoecaria agallocha* pure patch. The developed higher-order model shows higher accuracy than the linear spectral unmixing model, as it considers the interaction between the same species. The pixel value of a 30m × 30m pure mangrove species patch represents the reflectance of that particular mangrove species. When linear interactions are considered, the final pixel value is a representation of a single scattering of radiation from the target species to the sensor. The linear spectral unmixing model gives accurate results if the strands are arranged slightly apart side by side, as in a checkerboard. In such cases, the final reflectance value of a pixel represents a single reflection (radiation falling on a tree and getting reflected back and intercepted by the sensor). However, in pure and dense mangrove patches, the same type of mangrove species occurs close to each other and hence encounters multiple bounces of reflection between the closely spaced trees. In these cases, the final pixel value is a combination of single and multilevel reflections. As our intra-species interaction model considers linear as well as multiple-level reflections, it gives more accurate fractional abundance information for pixels representing a pure mangrove patch than a linear spectral unmixing model (Table 6.4).

Though higher-order models result in more accurate identification of endmembers and fractional abundance estimation, they have some limitations. The increase in order leads to an increase in computational complexity and processing time.

Moreover, in the case of mangrove species and their admixtures, the possibility of the presence of more than four species within a hyperspectral pixel area in the

TABLE 6.3

Comparative Analysis of Different Model Outputs in Relation to Mixed Mangrove Patches: Data Relates to Fractional Abundance Values and Root Mean Square Error at Geographic Coordinates 21o34.616′N; 88o16.512′E

| | Fractional Abundance Values | | | | | |
Endmembers	Linear Spectral Unmixing Model	Nascimento's Bilinear Model	Higher Order Model of Second Order	Higher Order Model of Third Order	Higher Order Model of Fourth Order	Ground Truth Results
1 (*Excoecaria agallocha*)	0.6489	0.4156	0.2505	0.2392	0.2383	0.24
2 (*Avicennia alba*)	0.1552	0.0938	0.0557	0.0584	0.0582	0.06
6 (*Bruguiera cylindrica*)	0.0955	0.0575	0.0403	0.0380	0.0378	0.04
7 (*Avicennia officinalis*)	0.1004	0.0666	0.0372	0.0368	0.0367	0.04
1,1			0.3580	0.3282	0.3269	0.33
1,4		0.1347	0.0795	0.0802	0.0799	0.08
1,6		0.0826	0.0576	0.0521	0.0519	0.05
1,7		0.0957	0.0532	0.0505	0.0503	0.05
4,4			0.0177	0.0196	0.0195	0.02
4,6		0.0186	0.0128	0.0127	0.0127	0.01
4,7		0.0216	0.0118	0.0123	0.0123	0.01
6,6			0.0093	0.0083	0.0082	0.01
6,7		0.0132	0.0086	0.0080	0.0080	0.01
7,7			0.0079	0.0078	0.0077	0
1,4,6				0.0175	0.0174	0.02
1,4,7				0.0169	0.0169	0.02
1,6,7				0.0110	0.0110	0.01
4,6,7				0.0027	0.0027	0
1,4,6,7					0.0037	0
Root Mean Square Error	0.2472	0.0842	0.0306	0.0214	0.0109	

TABLE 6.4
Abundance Estimation and Root Mean Square Error Values of Pure Patch of
Excoecaria agallocha

Endmembers	Linear Spectral Unmixing	Higher-Order Model for Intra-Species Interaction of Second Order	Ground Truth Results
1	1.000	0.5000	0.5
1,1		0.5000	0.5
Root Mean Square Error	0.2673	0.1109	

Geographical Coordinate: 21o34′9.40″N; 88o17′10.68″E

area under study (Henry Island) is almost negligible. For the Sunderban mangrove forest, in particular, computational complexity up to fourth order is manageable. For a higher number of endmembers, the linearly separable endmembers need to be separated before the application of the non-linear spectral unmixing model on them.

7 Fuzzy Logic-Based Non-Linear Spectral Unmixing

7.1 FUZZY LOGIC-BASED NON-LINEAR SPECTRAL UNMIXING

Fuzzy logic-based non-linear spectral unmixing is a method that utilizes the principles of fuzzy logic to perform non-linear spectral unmixing of hyperspectral data. The fuzzy logic approach allows for the modeling of uncertainty and ambiguity in the data, which is often present in real-world scenarios.

The basic idea behind fuzzy logic-based unmixing is to use fuzzy sets to represent the endmembers and their corresponding abundances in the data. The fuzzy sets are defined by their membership functions, which describe the degree of membership of each pixel to each endmember. The membership functions are usually based on some similarity measure, such as the Euclidean distance or the Mahalanobis distance, between the pixel spectrum and the endmember spectrum.

Fuzzy C-means (FCM) algorithm is one of the most popular algorithms for fuzzy logic-based spectral unmixing. The FCM algorithm starts with an initial set of centroids, which are the representatives of the endmembers. Then, it iteratively updates the membership functions and the centroids until convergence is reached. The final membership functions are then used to estimate the abundance of the endmembers in each pixel.

Another popular algorithm for fuzzy logic-based unmixing is the possibilistic C-means (PCM) algorithm. The PCM algorithm is similar to the FCM algorithm, but it uses possibilistic membership functions instead of fuzzy membership functions. The possibilistic membership functions are based on the possibility theory, which is a generalization of probability theory that allows for the modeling of partial and uncertain information.

Fuzzy logic-based unmixing methods have been successfully applied in several fields, such as mineral mapping, vegetation mapping, and water quality monitoring. These methods have shown to be robust to noise and outliers in the data, and they can handle endmember variability and mixed pixels better than the traditional linear methods.

Fuzzy logic-based non-linear spectral unmixing is a powerful method for the analysis of hyperspectral data. It allows for the modeling of uncertainty and ambiguity in the data, and it can handle endmember variability and mixed pixels better than the traditional linear methods. The FCM and PCM algorithms are two popular algorithms

DOI: 10.1201/9781003432623-7

for fuzzy logic-based unmixing. These methods have been successfully applied in several fields and have shown to be robust to noise and outliers in the data.

7.2 FUZZY C-MEANS (FCM)

FCM is a clustering algorithm that assigns a degree of membership to each data point for each cluster. This algorithm can also be used for unmixing hyperspectral data, where the endmembers are modeled as fuzzy clusters. The main advantage of FCM-based unmixing is its ability to handle mixed pixels, which are pixels that contain more than one endmember.

The basic idea behind FCM-based unmixing is to model the endmembers as fuzzy clusters and use the FCM algorithm to estimate the membership degrees of each pixel to each cluster. The membership degrees are used as the fractional abundances of the endmembers in each pixel.

The FCM algorithm consists of two steps: the assignment step and the update step. In the assignment step, the membership degrees of each pixel to each cluster is computed. In the update step, the cluster centers and the fuzzy exponent are updated. The algorithm iterates between these two steps until a stopping criterion is met.

The mathematical formulation of the FCM algorithm for unmixing is as follows:

1. Initialize the membership degrees $U = \{u_{ij}\}$ where u_{ij} is the membership degree of the ith pixel to the jth cluster
2. Compute the cluster centers $V = \{v_j\}$ where v_j is the jth cluster center
3. Update the membership degrees using the following equation:

$$u_{ij} = ((\|x_i - v_j\|^2) / \mathrm{sum}((\|x_i - v_k\|^2)^{\wedge}(1/(1-m)))))^{\wedge}(-1)$$

4. Update the cluster centers using the following equation:

$$v_j = (\mathrm{sum}(u_{ij}^m * x_i)) / (\mathrm{sum}(u_{ij}^m))$$

5. Repeat steps 3 and 4 until convergence,

where x_i is the ith pixel, m is the fuzzy exponent, and $\|.\|$ is the Euclidean distance.

An example of FCM-based unmixing is the application of the algorithm to a hyperspectral dataset of vegetation. The dataset contains three endmembers: green vegetation, dry vegetation, and soil. The FCM algorithm is used to estimate the membership degrees of each pixel to each cluster. The results show that the algorithm can accurately estimate the fractional abundances of the endmembers in each pixel.

It is important to note that the FCM algorithm assumes that the endmembers are Gaussian distributed and that the noise is Gaussian distributed as well. However, in practice, the noise in hyperspectral data is often non-Gaussian, so the performance of the FCM algorithm may be affected.

One of the advantages of FCM-based unmixing is that it can handle the presence of noise and outliers in the data. Additionally, it can also handle the case of endmembers with similar spectral signatures. However, one of the limitations of FCM-based unmixing is that it requires a large number of iterations to converge.

The FCM-based unmixing method has been applied to various hyperspectral datasets and has shown promising results in several applications, including mineral mapping, vegetation classification, and target detection. For example, in a study by Li et al. (2013), the FCM-based unmixing method was applied to a hyperspectral dataset of a typical coal mine area in China. The results showed that the FCM-based unmixing method outperformed other linear and non-linear unmixing methods in terms of overall accuracy and Kappa coefficient.

In conclusion, FCM-based non-linear spectral unmixing is a powerful method for the analysis of hyperspectral data. Its ability to handle noise and outliers, as well as its ability to handle the case of endmembers with similar spectral signatures, make it a suitable method for various applications. However, its high computational cost must be taken into consideration when applying this method to large datasets.

7.3 POSSIBILISTIC C-MEANS (PCM)

PCM is another fuzzy logic-based unmixing method that can be used for non-linear spectral unmixing. It is an extension of the fuzzy C-means (FCM) algorithm, and it is designed to handle the presence of noise and outliers in the data.

The PCM algorithm uses a membership function to represent the degree of association between a data point and a cluster center. Unlike the FCM algorithm, which uses a crisp membership function, the PCM algorithm uses a possibility membership function. This allows the algorithm to handle uncertain or incomplete information in the data.

The mathematical formulation of the PCM algorithm is similar to that of the FCM algorithm. The objective function is to minimize the sum of the squared distances between the data points and the cluster centers, subject to the constraint that the sum of the possibility memberships of a data point is equal to one.

The algorithm starts with an initial set of cluster centers and then iteratively updates the cluster centers and the possibility memberships of the data points until the objective function converges.

One example of the application of PCM on a dataset is the unmixing of hyperspectral images of a forest. PCM algorithm is applied to classify different tree species in the forest by considering the reflectance spectra of the trees as mixed endmembers. The algorithm was able to classify the different tree species with a high degree of accuracy and robustness against noise and outliers.

In conclusion, possibilistic C-means (PCM) is a robust algorithm for non-linear unmixing that can handle uncertain and incomplete information in the data. It is beneficial for hyperspectral image analysis, where the presence of noise and outliers is familiar.

7.4 ENTROPY-BASED FUZZY C-MEANS UNMIXING

Another algorithm that can be used for non-linear spectral unmixing is entropy-based fuzzy unmixing. This method uses the concept of fuzzy sets to model the uncertain and imprecise nature of the endmembers in a mixture. The algorithm maximizes the information entropy of the endmembers while satisfying the non-negativity and sum-to-one constraints. The entropy-based fuzzy unmixing algorithm has been shown to perform well in scenarios where the endmembers are highly correlated or overlapping. A variation of this algorithm is the fuzzy-entropy-based unmixing algorithm, which uses a combination of fuzzy C-means and entropy maximization to estimate the endmember signatures and fractions. However, this method is less used in comparison to other methods as it is computationally expensive and not as robust.

The EFCM is an iterative algorithm, and at convergence, the entire image is segmented into different endmember regions. Complement of Gaussian distribution function is considered as distance and integrated within the entropy-based fuzzy C-means objective function.

The smaller value of the distance term represents the strong association of the candidate pixel with the corresponding cluster prototype.

The objective function for the EFCM algorithm is defined as follows:

$$J(M,V) = \sum_{k=1}^{c} \sum_{i=1}^{N} u_{ik}{}^{m} \left(1 - e^{\frac{-\left\| x_i - u_k \right\|^2}{2\sigma_k^2}} \right) + \sum_{i=1}^{N} U\left(x_i \right)$$

where $m>1$ is the fuzziness coefficient;

u_{ik} is the degree of membership of x_i in the cluster k such that;

$U(x_i)$ is the class uncertainty associated with the given pixel intensity x_i

where P_{ik} is the probability of the pixel intensity x_i to belong to the kth class

$$U\left(x_i \right) = -\sum_{K=1}^{C} P_{ik} \log P_{ik}$$

where P_{ik} is the probability of the pixel intensity x_i to belong to the kth class

$$P_{ik} = \frac{a_{ik}}{\sum_{J=1}^{C} a_{ij}} \Bigg| a_{ik} = e^{\frac{-\left\| xi - uk \right\|^2}{2\sigma_k^2}}$$

In these equations, μ_k and σ_k denote the mean and the standard deviation of the intensity values for the kth class represents the uncertainty associated with each pixel which is defined as the Shannon entropy. The cluster center, v_j and membership u_{ij}, which are defined in the following equations, respectively.

$$u_{ik} = \frac{1}{\sum_{j=1}^{C} \left(\frac{1 - e^{\frac{-\left\| x_i - u_k \right\|^2}{2\sigma_k^2}}}{1 - e^{\frac{-\left\| x_i - u_j \right\|^2}{2\sigma_k^2}}} \right)^{\frac{1}{m-1}}}$$

$$\mu_k = \frac{\sum_{i=1}^{N}\left\{\left(1+\ln P_{ik}\right)P_{ik}+u_{ik}^{m}a_{ik}\right\}x_i}{\sum_{i=1}^{N}\left\{\left(1+\ln P_{ik}\right)P_{ik}+u_{ik}^{m}a_{ik}\right\}}$$

7.5 SPATIAL FUZZY-BASED UNMIXING

Spatial fuzzy unmixing is a method that combines fuzzy logic and spatial information to perform non-linear spectral unmixing. The method uses spatial information to improve the accuracy of the unmixing process by incorporating spatial constraints and taking into account the spatial correlation among pixels. The basic idea behind this method is to use a spatial fuzzy membership function to represent the degree of membership of each pixel to each endmember class. The spatial fuzzy membership function is calculated based on the spatial information of the pixel, such as its distance from the mean of each class and the degree of similarity between the pixel's spectral signature and the spectral signature of each class. The spatial fuzzy membership function is then used to estimate the fractional abundances of each endmember class in each pixel. Spatial fuzzy unmixing has been applied to various remote sensing applications, such as land cover classification, mineral mapping, and vegetation mapping, and has shown promising results.

One of the important characteristics of an image is that neighboring pixels are highly correlated and the probability that they belong to the same cluster is great. This spatial relationship is important in clustering, but it is not utilized in a standard FCM algorithm.

To exploit the spatial information, a spatial function is defined as

$$h_{ij} = \sum_{k \in NB(x_j)} u_{ik}$$

where $NB(x_j)$ represents a square window centered on pixel x_j in the spatial domain. A 5×5 window was used throughout this work. Just like the membership function, the spatial function h_{ij} represents the probability that pixel x_j belongs to ith cluster. The spatial function of a pixel for a cluster is large if the majority of its neighborhood belongs to the same clusters.

The spatial function is incorporated into the membership function as follows:

$$u'_{ij} = \frac{u_{ij}^{p} h_{ij}^{q}}{\sum_{k=1}^{c} u_{kj}^{p} h_{kj}^{q}}$$

where p and q are parameters to control the relative importance of both functions.

7.6 APPLICATIONS OF FUZZY-BASED NON-LINEAR UNMIXING

One application of non-linear unmixing algorithms, such as the ones based on fuzzy logic, is mineral mapping. For example, the fuzzy C-means (FCM) algorithm has been used to identify and map minerals in hyperspectral imagery.

In one study, FCM was used to unmix the reflectance spectra of a hyperspectral dataset acquired over a mineral deposit, and the resulting endmembers were then matched to reference spectra of known minerals. The method successfully mapped the distribution of several minerals, including clay minerals, iron oxides, and sulfates.

Another study used a modified version of PCM called possibilistic C-means (PCM) for mineral mapping of hyperspectral data. The PCM algorithm was able to accurately identify and map the presence of minerals, such as hematite, goethite, and kaolinite in the data.

One example of a mineral application is the identification and mapping of mineral deposits in hyperspectral data. These deposits are often composed of multiple minerals, and traditional linear unmixing methods may not accurately estimate each mineral's fractional abundance. By using FCM or PCM, we can incorporate fuzzy membership values for each pixel, which allows for a more accurate estimation of the fractional abundances of each mineral in the deposit.

Another example is the detection of hydrated minerals in hyperspectral data. Hydrated minerals, such as clay minerals, have unique spectral characteristics that can be used to identify them in hyperspectral data. However, these minerals often occur in a mixed state with other minerals, making it difficult to accurately estimate their fractional abundance using linear unmixing methods. By using FCM or PCM, we can take into account the overlapping spectral signatures of the minerals and estimate the fractional abundance of the hydrated mineral more accurately.

To obtain the results from these applications, the hyperspectral data must first be preprocessed to remove noise and atmospheric effects. Then, the FCM or PCM algorithm can be applied to the data to estimate the fractional abundances of the minerals. The results can be visualized in the form of maps that show the distribution and abundance of each mineral in the study area.

In terms of analyzing the results, it is important to evaluate the accuracy of the estimated fractional abundances. This can be done by comparing the results to ground truth data, if available, or by using statistical measures such as the coefficient of determination (R^2) or the root mean square error (RMSE). Additionally, it is essential to consider the computational efficiency of the algorithm and the number of endmembers required for proper unmixing.

Case Study-Based Results

In this section, the fuzzy-based methods are thoroughly compared for hyperspectral datasets such as Samson. The Samson dataset was captured in 156 spectral channels with spectral resolution 3.13 nm, ranging from 401 nm to 889 nm and contains three major endmembers. The qualitative analysis has been done for a scene of (95×95) pixels in the Samson dataset in Figure 7.1. The first column represents original image (band 96), the first row depicts water, the second row depicts vegetation, and the third row depicts soil. For EFCM, FCM, sFCM, the fourth row depicts the fractional abundance images for three endmembers.

A comparative analysis of non-linear-spectral unmixing models has been given in Table 7.1 in which a comparison has been done using non-linear spectral unmixing models applied on the Samson dataset. The abundance images of endmembers' soil, vegetation, water, vegetation-water interface in the dataset have been displayed in Figures 7.2–7.5, respectively.

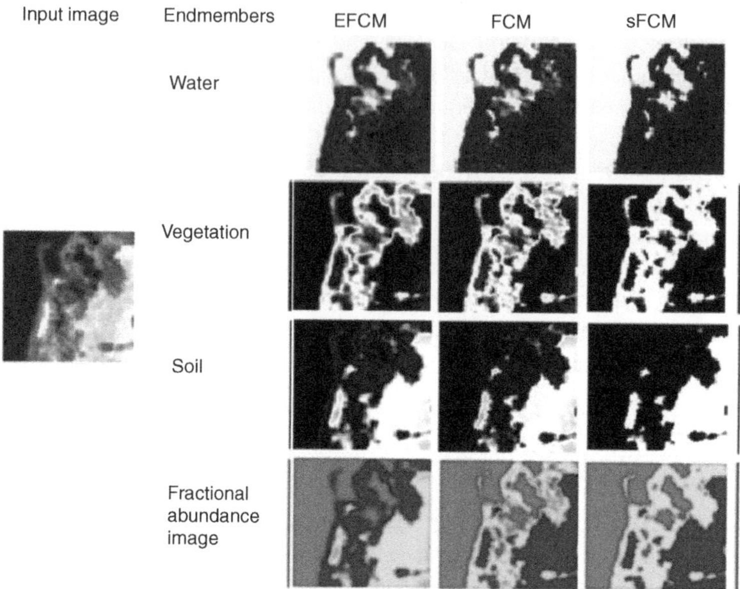

FIGURE 7.1   Qualitative results for the Samson dataset.

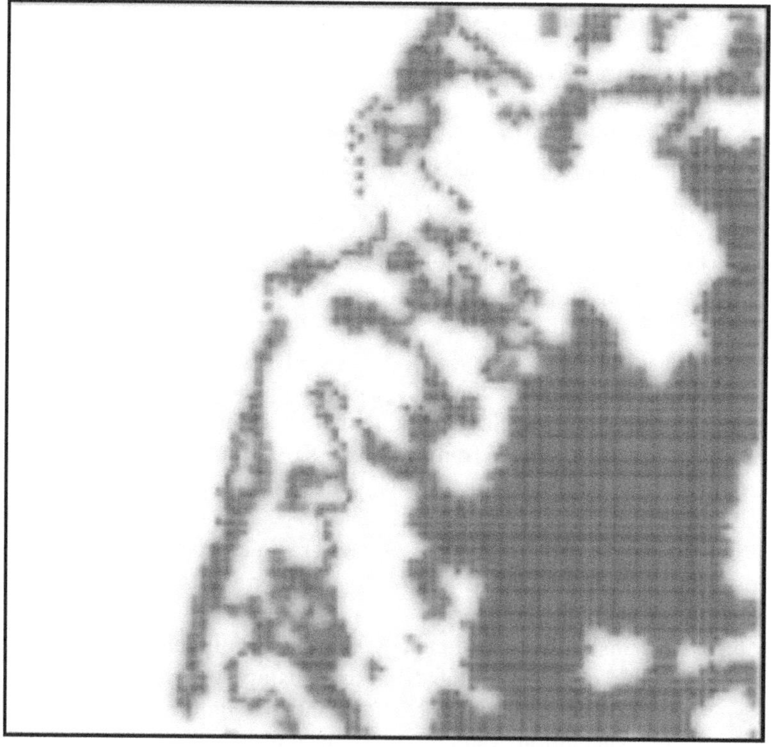

FIGURE 7.2   Abundance image of the soil endmember.

TABLE 7.1
Comparative Analysis of Non-Linear-Spectral Unmixing Models

Coordinate Considered	Method	Abundance Values of Respective Endmembers					
(13,3)		Soil	Vegetation	Water	Soil–vegetation	Soil–water	Vegetation–water
	Ground Truth	0	0	1	–	–	–
	Spatial FCM	0	0.13	0.62	0	–	0.2395
	FCM	0	0.11	0.55	0	–	0.3295
	Nascimento's model	0	0.0463	0.9534	0	0.0003	0
	Fan's model	0	0	0.9976.	0	0.0023	0
(67,53)		Soil	Vegetation	Water	Soil–vegetation	Soil–water	Vegetation–water
	Ground truth	0.7974	0.1966	0.0061	–	–	–
	Spatial FCM	0.2345	0.0184	0.0107	0.6574	–	0.079
	FCM	0.1229	0.0076	0.0105	0.8538	–	0.0051
	Nascimento's model	0.6674	0.1735	0.1574	0	0	0.0020
	Fan's model	0.99	0.002	0	0.001	0	0
(88,70)		Soil	Vegetation	Water	Soil–vegetation	Soil–water	Vegetation–water
	Ground truth	0.7760	0.2065	0.0174	–	–	–
	Spatial FCM	0.3172	0.0961	0.0023	0.5844		0
	FCM	0.1269	0.0085	0.0003	0.8538		0
	Nascimento's model	0.5941	0.0939	0.3101	0	0	0.0026
	Fan's model	0.99	0	0			
(90,95)		Soil	Vegetation	Water	Soil–vegetation	Soil–water	Vegetation–water
	Ground truth	0.9531	0	0.0469	–	–	–
	Spatial FCM	0.6897	0.0078	0.0157	0.2844	–	0.0024
	FCM	0.6385	0.0066	0.0009	0.3532	–	0.0008
	Nascimento's model	0.4981	0	0.5001	0	–	0.0041
	Fan's model	0.9997	0	0	0	0	0

FIGURE 7.3 Abundance image of the vegetation endmember.

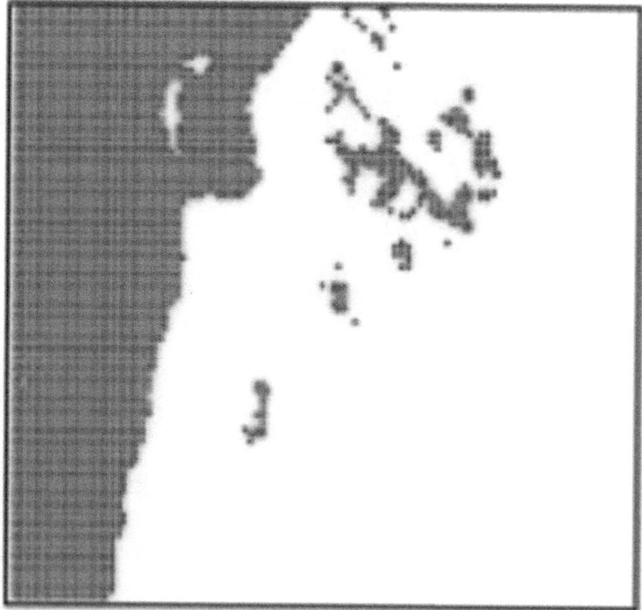

FIGURE 7.4 Abundance image of the water endmember.

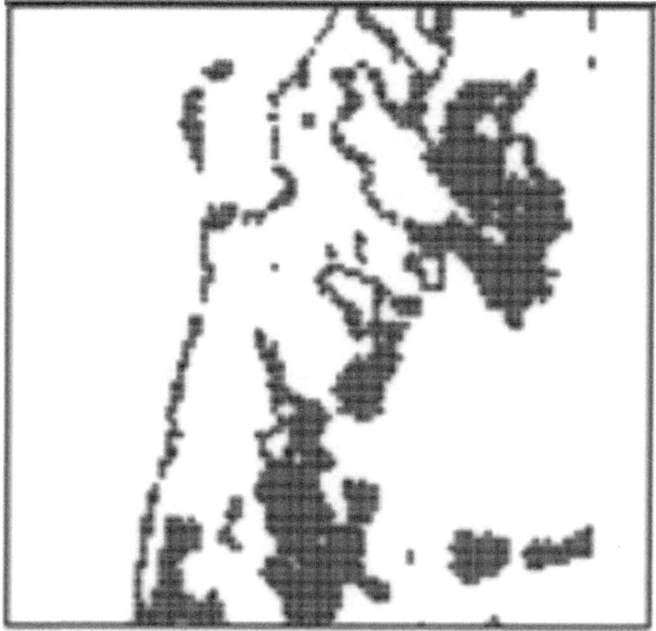

FIGURE 7.5 Abundance image of the vegetation–water interface endmember.

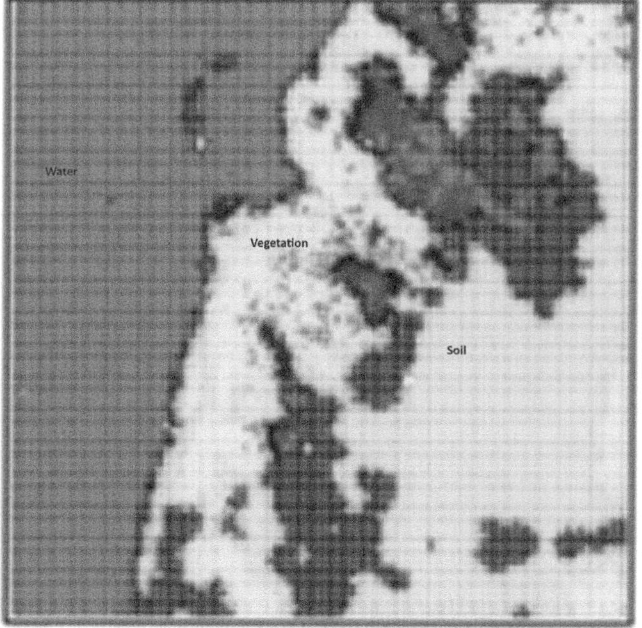

FIGURE 7.6 Classified abundance image.

In conclusion, fuzzy logic-based non-linear spectral unmixing algorithms like FCM and PCM have the ability to provide more accurate fractional abundance estimates of minerals in hyperspectral data compared to linear unmixing methods. These algorithms have been successfully applied in mineral exploration and other fields and have the potential for further research and development.

8 Machine Learning Models for Classification of Hyperspectral Data

8.1 MACHINE LEARNING MODELS FOR CLASSIFICATION OF HYPERSPECTRAL DATA

Classification of hyperspectral data is a critical task in remote sensing and environmental monitoring. Hyperspectral data refers to a type of data that captures the reflectance or emission spectrum of a scene or object over a wide range of wavelengths. This type of data is rich in information, but also high-dimensional and often highly correlated, which makes its analysis challenging. Machine learning models have become a popular tool for hyperspectral data classification, as they can effectively capture complex relationships between the spectral and spatial information in the data.

In this chapter, we will explore several machine learning models that are commonly used for hyperspectral data classification. We will start by introducing the basic concepts of machine learning and its applications to hyperspectral data. Then, we will delve into the different models and their underlying mechanisms, including support vector machines (SVMs), decision trees (DTs), random forests (RFs), artificial neural networks (ANNs), and deep learning models.

Support vector machines (SVMs) are a type of machine learning model that can be used for both classification and regression tasks. They are based on the idea of finding a hyperplane in the feature space that separates the classes with the largest margin. SVMs are powerful models that can handle non-linear relationships between the features and the classes by transforming the data into a higher-dimensional space and finding a separating hyperplane in that space.

Decision trees (DTs) and random forests (RFs) are two popular models for hyperspectral data classification. DTs are simple tree-based models that recursively split the feature space into smaller regions based on the class label. The final decision of a DT is based on the class label of the region in which a sample falls. RFs are an extension of DTs that create an ensemble of DTs and combine their predictions to make a final decision.

Artificial neural networks (ANNs) are a type of machine learning model that is inspired by the structure and function of the human brain. ANNs consist of multiple interconnected nodes, or neurons, that process information and make predictions based on the input data. ANNs can be used for both classification and regression tasks, and they can handle non-linear relationships between the features and the classes.

DOI: 10.1201/9781003432623-8

Deep learning models are a type of artificial neural network that have multiple layers and can learn hierarchical representations of the data. Deep learning models are effective in a wide range of tasks, including hyperspectral data classification. Convolutional neural networks (CNNs) and recurrent neural networks (RNNs) are two popular deep learning models that have been applied to hyperspectral data classification.

8.2 CLASSIFICATION MODELS

Classification of hyperspectral data is an important task in remote sensing and image analysis. Hyperspectral data refers to a large number of contiguous spectral bands that capture the unique reflectance properties of different materials in a scene. Machine learning algorithms are widely used in hyperspectral classification due to their ability to handle large amounts of data and perform complex operations.

There are several machine learning models used for hyperspectral data classification, including the following:

(i) Support vector machines (SVMs)

SVMs are a type of machine learning algorithm that are used for classification and regression analysis. In the context of hyperspectral data, SVM can be applied for different tasks, such as land cover classification, target detection, and spectral unmixing.

Land cover classification: SVM can be used for land cover classification by mapping the hyperspectral data to different classes, such as urban, agriculture, and forest, etc. The algorithm maps the high-dimensional hyperspectral data into a lower-dimensional feature space, where the decision boundary between classes can be defined. An SVM classifier can be trained by finding the optimal hyperplane that separates the data into their respective classes. The SVM algorithm can be applied to hyperspectral data in both two-class and multiclass classification scenarios.

Target detection: SVM can also be used for target detection in hyperspectral images. The target pixels can be considered as positive samples, and the non-target pixels as negative samples. The SVM algorithm can be trained using these samples to distinguish between targets and non-targets. Once trained, the SVM classifier can be applied to the hyperspectral data to identify target pixels.

Spectral unmixing: SVM can also be used for spectral unmixing, which is the process of separating mixed spectral signatures into their constituent materials. The SVM algorithm can be trained using known spectral signatures to separate mixed pixels into their respective materials. SVM-based spectral unmixing algorithms are effective in separating mixed pixels in hyperspectral data.

The algorithm can handle high-dimensional data, making it well-suited for hyperspectral data analysis. The algorithm's ability to learn complex non-linear decision boundaries and perform both two-class and multiclass classification makes it a versatile tool for hyperspectral data analysis.

(ii) Decision Trees and Random Forests

Decision trees and random forests are two popular machine learning algorithms often used for hyperspectral data analysis.

Decision Trees

A decision tree is a tree-based model that makes predictions based on a set of rules learned from the input data. The tree structure allows the model to iteratively divide the feature space into smaller sub-spaces, making predictions at each node based on the conditions specified in that node. In the case of hyperspectral data, the decision tree would consider different wavelengths of light as features and use them to determine the presence of different materials in an image.

For example, a decision tree could be trained to identify buildings, vegetation, and water bodies based on their reflectance characteristics in different spectral bands in a hyperspectral image of an urban area. At each node of the tree, a decision could be made based on the intensity of a specific wavelength or combination of wavelengths. The next node would consider a different set of wavelengths until a final prediction is made at the leaf node.

Random Forests

Random forests are an extension of decision trees, where multiple trees are grown and combined to produce a more robust model. The algorithm randomly samples the feature space, trains multiple trees independently, and then aggregates the results of all trees to make a final prediction. This combination of multiple decision trees can help reduce overfitting and increase the model's accuracy.

In the context of hyperspectral data, random forests can be used to perform a similar task as decision trees. For example, a random forest model could be trained to classify the different materials in an urban area based on their reflectance properties in different wavelengths. The model would sample the feature space multiple times, creating multiple decision trees, and then combine the results to produce a final classification map.

Overall, both decision trees and random forests are popular algorithms for hyperspectral data analysis due to their ease of use, interpretability, and ability to handle high-dimensional data. However, as with any machine learning model, it is important to carefully evaluate these models' performance and consider the trade-off between accuracy and interpretability.

(iii) Artificial Neural Networks (ANNs)

ANNs are a class of machine learning models inspired by the human brain's structure and function. They are composed of interconnected nodes, or artificial neurons, that process information and learn from data. In hyperspectral data analysis, ANNs have been applied in several ways to perform tasks, such as classification, feature extraction, and data dimensionality reduction.

One example of the application of ANNs on hyperspectral data is for the classification of land cover types. Hyperspectral data collected by a remote sensing instrument is used to train an ANN to recognize different land cover types, such as forests, crops, deserts, and urban areas. The training process involves presenting the network with a set of labeled examples, where each example consists of a hyperspectral feature vector and a corresponding land cover label. The network uses this information to learn the relationship between the hyperspectral data and the land cover labels, and this information is stored in the connection weights between the neurons.

Once the network has been trained, it can be used to classify new hyperspectral data it has not seen before. Making predictions with an ANN involves passing a hyperspectral feature vector through the network and obtaining the network's output, which is a set of class probabilities. The class with the highest probability is then selected as the final prediction.

Another example of the application of ANNs on hyperspectral data is for the task of data dimensionality reduction. In this case, an ANN is used to learn a lower-dimensional representation of the hyperspectral data while preserving the information content as much as possible. This is useful because hyperspectral data can have hundreds or thousands of spectral bands, and processing such high-dimensional data can be computationally expensive. By reducing the dimensionality of the data, it becomes easier and more efficient to perform further processing and analysis.

To perform data dimensionality reduction with an ANN, a network with an input layer, one or more hidden layers, and an output layer is trained using a large data set of hyperspectral data. The input layer receives the raw hyperspectral data, while the hidden layers learn to extract a lower-dimensional representation of the data. The output layer provides a lower-dimensional representation, which can be used for further processing. The training process involves adjusting the connection weights between the neurons in the network to minimize the difference between the output of the network and the original hyperspectral data.

ANNs are a powerful tool for analyzing hyperspectral data and have been applied in various tasks, including classification, feature extraction, and data dimensionality reduction. By learning from the data and making predictions based on that learning, ANNs can provide valuable insights into the information in hyperspectral data.

(iv) *K*-nearest neighbors (KNN)

KNN is a simple, non-parametric, instance-based, supervised machine learning algorithm for classification and regression. The basic idea behind KNN is to classify an unknown sample based on the majority vote of its k-nearest neighbors in the training set. The k-nearest neighbors are determined based on a distance metric, such as Euclidean distance, Mahalanobis distance, or other distance measures.

In hyperspectral data, KNN has been applied for several tasks, such as classifying pixels into different land use and land cover classes, target detection, and unmixing. In hyperspectral image classification, KNN can be used to determine the class label of each pixel based on its k-nearest neighbors in the training set. For example, in a hyperspectral data set of an urban area, the pixels can be classified into different land use and land cover classes, such as buildings, roads, vegetation, and water bodies.

For target detection in hyperspectral data, KNN can be used to identify pixels that belong to a target in the scene. In this case, the pixels are first classified into different land use, and land cover classes using KNN, and pixels belonging to a target are selected based on the target's signature in the hyperspectral data. For example, in a hyperspectral data set of a forest, KNN can be used to classify pixels into vegetation and non-vegetation classes. Then pixels that belong to a specific tree species can be selected based on their unique spectral signature in the hyperspectral data.

In hyperspectral unmixing, KNN can estimate the abundances of different endmembers in a pixel based on the k-nearest neighbors in the training set. For example, in a hyperspectral data set of a soil sample, KNN can be used to estimate the abundances of different soil minerals in each pixel based on the mineral signatures in the hyperspectral data.

Overall, KNN is a simple and effective algorithm that can be applied to various hyperspectral data analysis tasks. Its main advantage is its simplicity, but its accuracy may suffer from the curse of dimensionality, which means that the algorithm may perform poorly when the number of features is large compared to the number of samples.

(v) Convolutional neural networks (ConvNets or CNNs)

CNNs are deep learning neural networks primarily used in image classification and computer vision tasks. However, their application has been extended to hyperspectral data in recent years.

Hyperspectral data is characterized by high-dimensional spectral information, where each pixel in an image has a reflectance spectrum with hundreds or thousands of bands. This data type is beneficial in various remote sensing applications, such as mineral mapping, crop classification, and land-use mapping.

The application of ConvNets to hyperspectral data involves using a ConvNet architecture to extract features from the data and make predictions about the class of each pixel in the image. ConvNets are particularly well suited for this task because of their ability to learn local and hierarchical representations of the input data through convolutional and pooling layers.

One common approach is to use a 2D ConvNet, where each band in the hyperspectral image is treated as a separate channel in a 2D image. This type of ConvNet can be trained to recognize patterns and relationships between the different bands in the hyperspectral image, which can be useful for classifying pixels into different classes.

Another approach is to use a 3D ConvNet, which considers the spatial and spectral relationships between the pixels in the hyperspectral image. In this case, ConvNet is trained to learn both local and global representations of the input data by using 3D filters that operate on the spatial and spectral dimensions of the data.

One example of ConvNet application on hyperspectral data is using 3D ConvNets for land-use classification. In this case, a 3D ConvNet is trained on a data set of hyperspectral images of various land-use types, such as forests, urban areas, and croplands. The ConvNet is then used to make predictions about the class of each pixel in a new hyperspectral image, providing a high-resolution classification map of the land-use types in the area.

Another example is using 2D ConvNets for mineral mapping in hyperspectral data. In this case, a 2D ConvNet is trained on a data set of hyperspectral images of different minerals, such as iron ore, copper, and gold. The ConvNet is then used to make predictions about the mineral type of each pixel in a new hyperspectral image, providing a high-resolution mineral map of the area.

Convolutional neural networks are a powerful tool for hyperspectral data analysis and have been applied successfully in various tasks, such as hyperspectral image classification, dimensionality reduction, and feature extraction. ConvNets are able to automatically learn hierarchical feature representations and retain the spatial information inherent in the data, making them well-suited for hyperspectral data analysis.

8.3 PREDICTION MODELS

In hyperspectral data analysis, prediction models refer to machine learning algorithms that aim to predict a continuous target variable based on a set of input features. Unlike classification models, which predict categorical labels, prediction models predict numerical values.

(i) *Linear regression*: linear regression is a simple and widely used prediction model that assumes a linear relationship between the input features and the target variable. The model estimates the coefficients of the linear equation that best fits the training data. The linear regression model can be applied to hyperspectral data for applications, such as estimating the abundance of a mineral in the ground or predicting the water content of crops from hyperspectral images.

(ii) *Non-linear regression*: non-linear regression is an extension of linear regression that allows for more complex relationships between the input features and target variable. It involves the use of non-linear functions to fit the data, and is often used for hyperspectral data where the relationship between the input features and target variable is not linear.

(iii) *Support vector regression (SVR)*: SVR is a regression algorithm that uses support vectors to fit a regression model. Unlike linear regression, SVR is not limited to linear relationships between the input features and target variable, and can handle non-linear relationships by transforming the data into a higher-dimensional space. SVR can be applied to hyperspectral data to predict variables, such as soil moisture content or vegetation cover.

(iv) *Artificial neural networks (ANNs)*: ANNs are a type of machine learning algorithm that are inspired by the structure and functioning of the human brain. ANNs can be used for regression problems by using multiple hidden layers to model the complex relationships between the input features and target variable. ANNs are commonly applied to hyperspectral data for prediction tasks, such as estimating atmospheric parameters or predicting soil properties.

(v) *Random forest regression*: random forest regression is an ensemble method that combines multiple decision trees to make predictions. The algorithm generates multiple decision trees using bootstrapped samples of the training data, and the final prediction is made by averaging the predictions of the individual trees. Random forest regression is commonly applied to hyperspectral data to predict variables, such as vegetation cover or land surface temperature.

These are some of the most commonly used prediction models for hyperspectral data analysis. The choice of model will depend on the specific problem being addressed, the characteristics of the data, and the computational resources available.

8.3.1 Deep Learning

Deep learning models are a sub-field of machine learning that use artificial neural networks with multiple layers to learn patterns and relationships in data. They have been widely used in various domains, including computer vision, speech recognition, and natural language processing, and have been shown to achieve state-of-the-art results in many tasks. In hyperspectral data analysis, deep learning models can be used for various applications, such as classification, dimensionality reduction, feature extraction, and end-to-end hyperspectral image processing.

One popular deep learning model applied to hyperspectral data is convolutional neural networks (CNNs). A CNN is designed to learn spatial patterns in data by convolving the input data with filters and pooling the results to reduce the spatial dimensions while maintaining the most important information. In hyperspectral data analysis, CNNs can be used to classify the pixels of a hyperspectral image into different classes based on their spectral and spatial characteristics. For example, a CNN can be trained on a large data set of hyperspectral images to learn the patterns and relationships between different classes, such as vegetation, urban areas, and bare soil. After training, the CNN can be used to classify new hyperspectral images into the same classes based on their spectral and spatial features.

Another deep learning model applied to hyperspectral data is autoencoders. Autoencoders are neural networks that aim to learn a compact representation of the input data by encoding it into a lower-dimensional space and then decoding it back into the original space. In hyperspectral data analysis, autoencoders can be used for dimensionality reduction, which can reduce the computational costs and improve the accuracy of subsequent analysis steps. For example, an autoencoder can be trained on a large data set of hyperspectral images to learn a compact representation of the data while maintaining the most important information. After training, the autoencoder can be used to reduce the dimensionality of new hyperspectral images, making subsequent analysis steps more efficient and accurate.

Finally, deep belief networks (DBNs) and recurrent neural networks (RNNs) are also applied to hyperspectral data. DBNs are deep learning models that consist of multiple layers of restricted Boltzmann machines (RBMs), which are generative models that can learn a compact representation of the data. In hyperspectral data analysis, DBNs can be used for feature extraction and dimensionality reduction. RNNs are deep learning models that are designed to learn patterns in sequential data. In hyperspectral data analysis, RNNs can be used to model the temporal relationships between the pixels of a hyperspectral image sequence, which can provide valuable information for tasks, such as change detection and monitoring.

Deep learning models have been shown to be effective for various tasks in hyperspectral data analysis, including classification, dimensionality reduction, feature extraction, and end-to-end hyperspectral image processing. These models have the ability to learn complex patterns and relationships in the data and have been shown to achieve state-of-the-art results in many tasks.

8.4 MERITS AND DEMERITS OF THE MODELS THROUGH EXPERIMENTAL ANALYSIS

Experimental analysis is a crucial aspect of evaluating the performance of various machine learning models on hyperspectral data. Merits and demerits of different models can be determined through such an analysis. This section will discuss the evaluation criteria and results of different classification, prediction, and deep learning models applied on hyperspectral data.

Classification Models

Merits

Simple and easy to implement: classification models like k-nearest neighbors (KNN), decision trees, and random forests are simple to implement and understand. They do not require extensive mathematical knowledge.

Fast and efficient: classification models are fast and efficient in terms of processing time and memory usage. This makes them ideal for large datasets.

High accuracy: when applied correctly, classification models can produce high accuracy results, making them ideal for hyperspectral data analysis.

Demerits

Overfitting: classification models can overfit the data, producing results that are too specific to the training data set and not generalizable to new data. This can result in poor performance on new data.

Lack of interpretability: it is often difficult to understand why a model made a particular prediction, making it difficult to make decisions based on the results.

Prediction Models

Merits

Good at extrapolation: prediction models like linear regression and support vector regression (SVR) are good at extrapolating beyond the range of the training data. This makes them useful for hyperspectral data, where data is collected at various points in time or space.

Flexible: prediction models can be applied to many problems and are highly customizable. They can be tailored to the specific needs of the problem at hand.

Demerits

Limited ability to capture non-linear relationships: predictive models cannot always capture non-linear relationships in the data, which can result in a poor fit to the data.

Computational cost: predictive models can be computationally expensive, requiring more processing power and memory than other methods.

Deep Learning Models

Merits

Good at capturing complex relationships: deep learning models like artificial neural networks (ANNs) and convolutional neural networks (CNNs) can capture complex relationships in the data. This makes them ideal for hyperspectral data, where the relationships between the spectral bands and the target variables can be complex.

High accuracy: deep learning models have been shown to achieve high accuracy results in various domains, making them ideal for hyperspectral data analysis.

Demerits

High computational cost: deep learning models can be computationally expensive, requiring high processing power and memory to train.

Overfitting: deep learning models can easily overfit the data, producing results that are too specific to the training data set and not generalizable to new data.

Black box: deep learning models are often referred to as 'black boxes' because it is often difficult to understand why they made a particular prediction. This can make it difficult to make decisions based on the results.

Each type of machine learning model has its own merits and demerits. The choice of model will depend on the specific requirements of the problem at hand, such as the size and complexity of the data, the computational resources available, and the desired accuracy of the results.

8.5 ROLE OF MACHINE/DEEP LEARNING IN SPATIAL BIG DATA ANALYTICS

Machine/deep learning (ML/DL) has played a significant role in the field of spatial big data analytics. Spatial big data refers to large data sets with a geographical or spatial component, such as satellite imagery, GPS tracks, and geospatial information from social media. The increasing amount of spatial big data has challenged traditional methods of data analysis and management. The traditional methods are often time-consuming, limited in their ability to handle large volumes of data, and unable to extract meaningful information.

The ML/DL approaches have provided a solution to these challenges by automating the analysis process, handling large volumes of data, and providing accurate results. These methods have the ability to learn patterns and relationships in the data without the need for manual intervention, which makes them highly suitable for analyzing spatial big data.

One of the main applications of ML/DL in spatial big data analytics is image classification. This is the process of automatically categorizing image data into different classes based on features, such as texture, color, and shape. Deep convolutional neural networks (DCNNs) have proven to be highly effective for this task, with a wide range of pretrained models available for use.

Another important application of ML/DL in spatial big data analytics is land cover and land-use classification. This involves mapping the types of land cover and land use in a given area. Land cover refers to the physical cover of the land, such as forests, grasslands, and wetlands. Land use refers to the way in which land is used, such as agriculture, urban areas, and parks. The use of ML/DL algorithms has significantly improved the accuracy of land cover and land-use classification, particularly in large and complex data sets.

Another application of ML/DL in spatial big data analytics is change detection. This involves identifying changes in the earth's surface over time, such as urban expansion, deforestation, and land-use changes. The use of ML/DL algorithms has improved the accuracy and efficiency of change detection, particularly in large and complex data sets.

ML/DL has played a significant role in spatial big data analytics by automating the analysis process, handling large volumes of data, and providing accurate results. The ability of these methods to learn patterns and relationships in the data has led to improved accuracy in tasks such as image classification, land cover and land-use classification, and change detection. As the amount of spatial big data continues to increase, the use of ML/DL approaches is likely to become even more important in spatial big data analytics.

CASE STUDY: EVALUATING THE PERFORMANCE OF A CONVOLUTIONAL NEURAL NETWORK (CNN) MODEL IN CLASSIFYING HYPERSPECTRAL IMAGES IN THE ANDAMAN AND NICOBAR ISLANDS

The Andaman and Nicobar Islands contain a significant area of mangroves, considered to be one of the most vulnerable coastlines in Southeast Asia (Figure 8.1). The authors of this study aim to better understand the environmental health of these ecosystems by using hyperspectral image classification through a deep learning model, a convolutional neural network (CNN).

Methodology

The performance of the USGS data set in two other regions, Henry Island and Lothian Island, was evaluated in two stages: preprocessing and deep learning-based classification. The region of interest (ROI) in the input image was detected after removing noise, improving the accuracy of predicted classes. The ROI process detects the boundaries

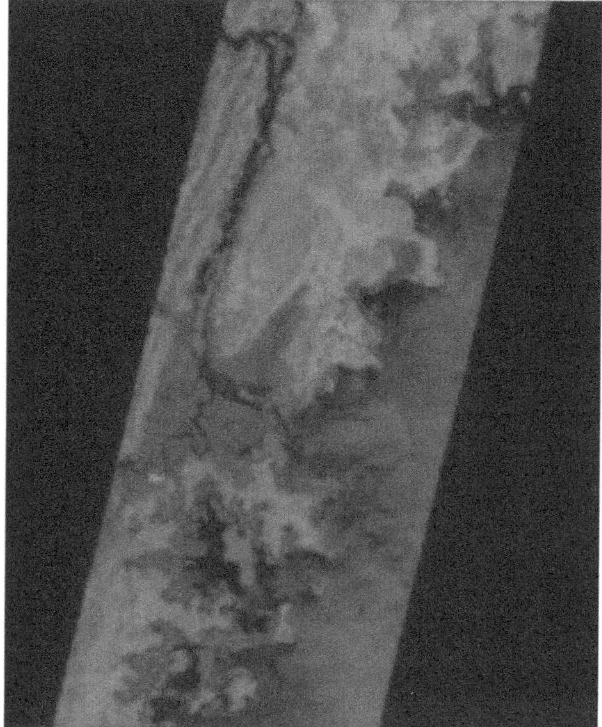

FIGURE 8.1   Hyperspectral data set of the Andaman & Nicobar Islands.

of each class to be classified and is efficient in predicting the coverage area of each class in the full image size. Figure 8.2 shows the major principal components of the Andaman data set after dimensionality reduction.

Figure 8.3 shows the reference spectra of mangroves and non-mangroves extracted to identify and compare different spectral signatures of the hyperspectral image. Figure 8.4 shows the spectral signature of different regions with standard benchmark spectra to ensure the existence of non-mangrove in ROI.

The deep CNN algorithm automatically classifies the input images by extracting significant features from the images. The algorithm learns the images and predicts their position and scale, improving the classification of images in future iterations. The number of layers in the deep CNN enhances the image for class prediction. These layers are responsible for extracting all features, converting pixels into vectors, weight estimation, mapping, and so on. The input images are processed in parallel with multiple neurons on each layer of the deep CNN.

Results and Discussion

After testing an image from each data set, a confusion matrix was determined to evaluate the classification performance. The confusion matrix computes accuracy, a significant metric that measures the efficiency of the proposed method. It is defined

Principal Component Bands of Data Cube

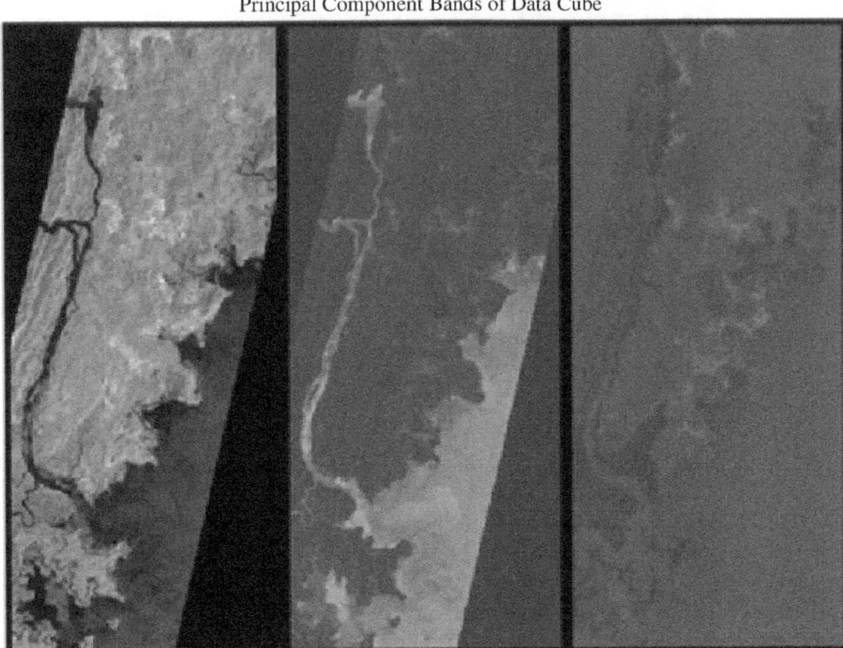

FIGURE 8.2 Major principal components of the Andaman data set after dimensionality reduction.

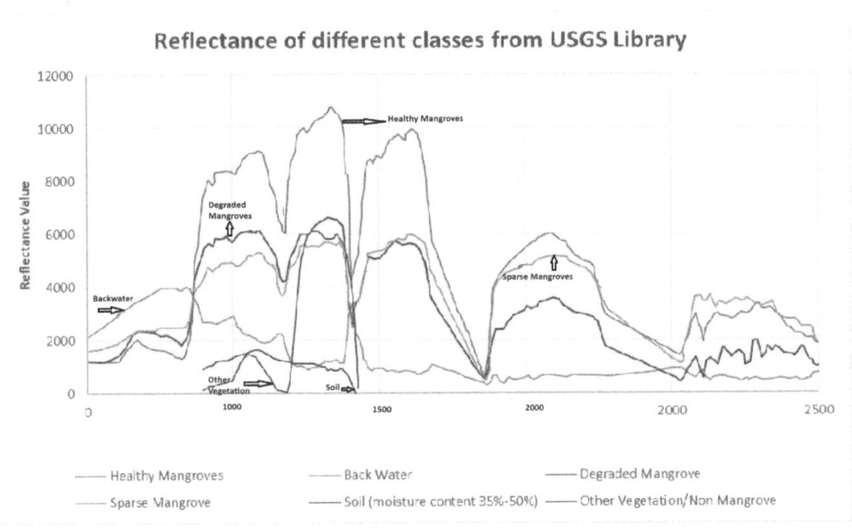

FIGURE 8.3 Reference spectra of mangroves and non-mangroves.

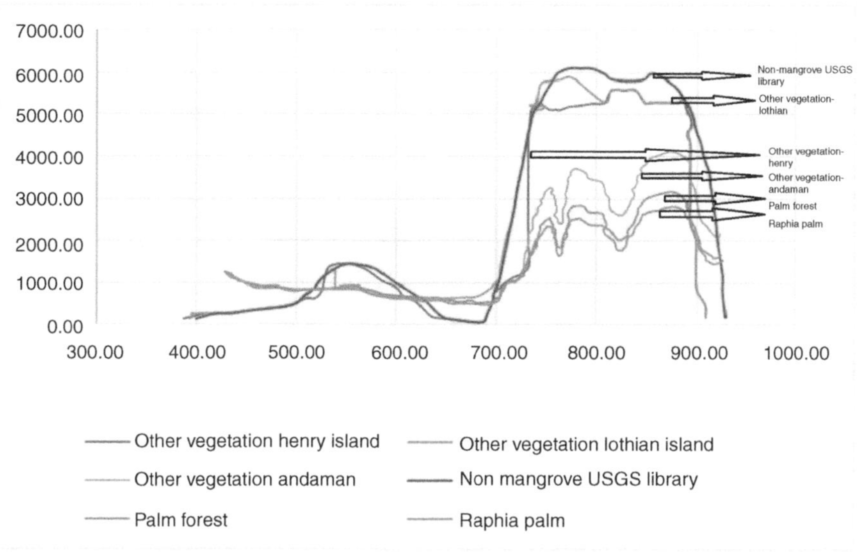

FIGURE 8.4　Spectral signature of different regions with standard benchmark spectra.

as the exactness in predicting classes in the appropriate categories. Figure 8.6 shows the confusion matrix evaluated for the USGS dataset in Andaman region, India, Saptamukhi Reserve Forest (Henry Island) Sundarbans, West Bengal, India, and Lothian Island, Sundarbans, West Bengal, India. The measured accuracy for both data sets in USGS was evaluated and is demonstrated in Figure 8.7. The USGS dataset in the Andaman region achieved higher accuracy than the Sundarbans region due to the limited coverage of classes compared to the Sundarbans region.

Precision is another critical metric used to measure the performance of the proposed algorithm. It is defined as the prediction of the exact classification quality and is also known as the positive predictive value. The precision of the USGS data set in the Andaman region was higher than in the Sundarbans region. The increase in precision denotes the proposed algorithm's ability to extract relevant results, i.e., accurate predictions of classes. Figure 8.5 shows the classification test results of the data sets of the three regions. Figure 8.7 depicts the comparison of accuracy and precision in different regions.

The confusion matrix also estimates true positive (TP), false positive (FP), false negative (FN), and true negative (TN). The accuracy and precision were calculated using the following mathematical formulations:

$$A = (TP + TN)/(TP + FP + TN + FN),$$

$$PS = TP/(TP + FP),$$

Different machine learning classifiers are tested on different regions, and the analysis of overall accuracy is measured to find the suitable model for hyperspectral

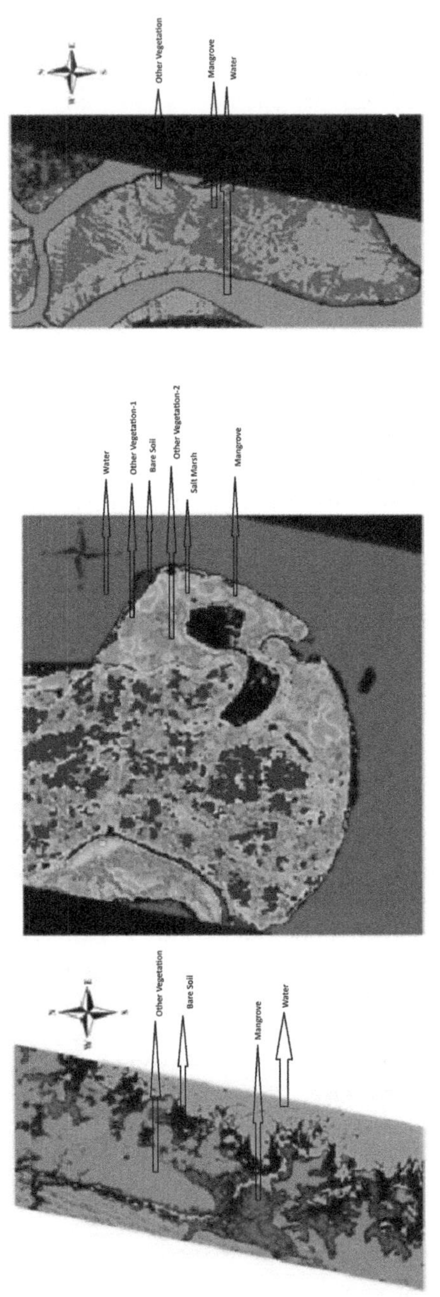

FIGURE 8.5 USGS data set classification test results: (a) Andaman Region, India; (b) Saptamukhi Reserve Forest (Henry Island) Sundarbans, West Bengal, India; and (c) Lothian Island, Sundarban, West Bengal, India.

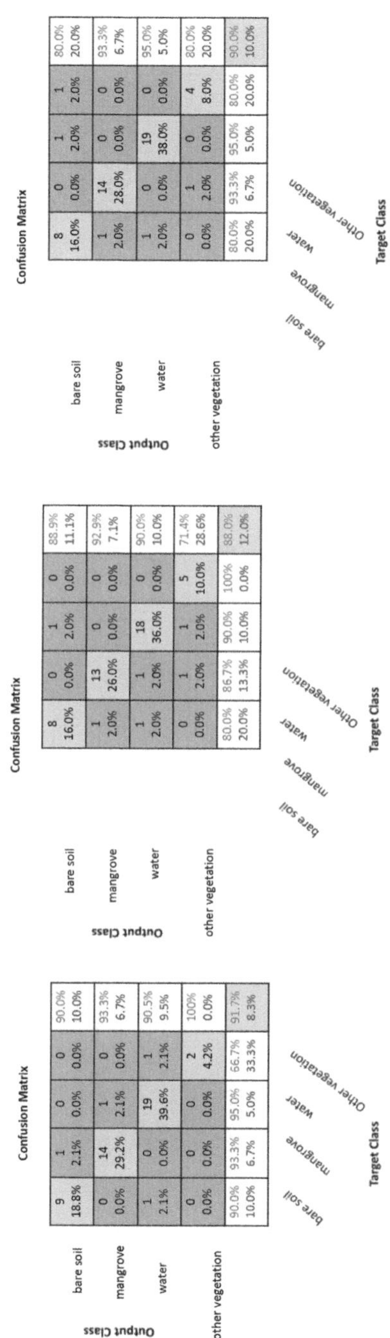

FIGURE 8.6 Confusion matrix evaluated for USGS dataset in (a) Andaman Region, India, (b) Saptamukhi Reserve Forest (Henry Island) Sundarbans, West Bengal, India, and (c) Lothian Island, Sundarbans, West Bengal, India.

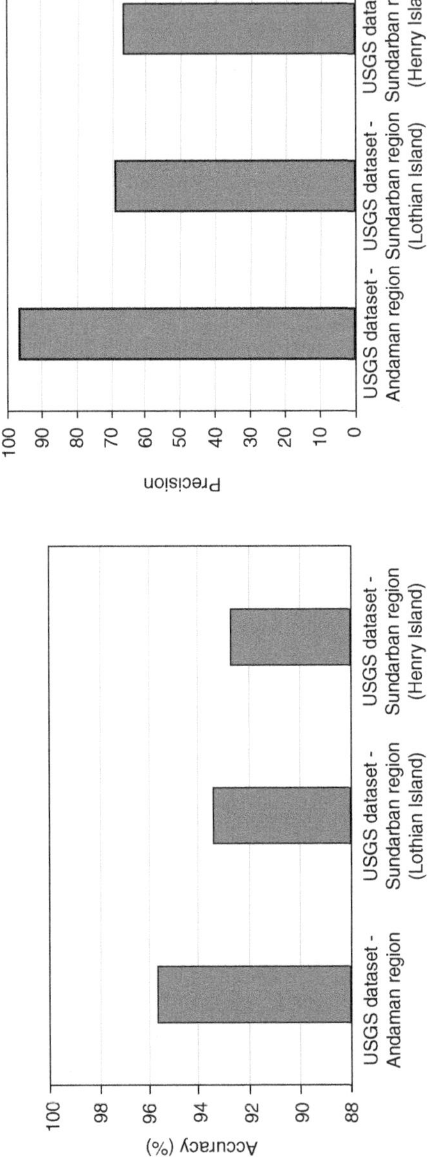

FIGURE 8.7 Comparison of accuracy and precision in different regions.

TABLE 8.1
Measured Classification Overall Accuracy Concerning Individual Classes under Consideration of Different Machine Learning Models

Class	Class Label	Andaman				Lothian				Henry			
		RF	SVM	CNN	3D DWT + CNN	RF	SVM	CNN	3D DWT + CNN	RF	SVM	CNN	3D DWT + CNN
1	Bare Soil	68.2	73.1	87.1	90.0	61.1	63.1	78.1	80.0	48.7	69.1	85.3	88.9
2	Mangrove	77.1	79.8	91.3	93.3	71.4	81.8	91.1	93.3	57.6	59.2	89.5	92.9
3	Water	56.1	60.4	80.2	90.5	43.1	60.4	93.2	95.0	43.1	59.7	87.4	90.0
4	Other Vegetation	47.3	65.2	99.1	100.0	57.2	64.2	78.1	80.0	49.3	55.9	67.8	71.4

TABLE 8.2
Measured Classification Speed of Different Machine Learning Models

Algorithm	Andaman		Lothian		Henry	
	Training Time (Sec)	Classification Speed (Sec)	Training Time (Sec)	Classification Speed (Sec)	Training Time (Sec)	Classification Speed (Sec)
Random Forest	24.59	2650	20.56	2650	16.48	2650
SVM	32.45	2370	26.41	2370	19.29	2370
CNN	46.89	2150	38.98	2150	36.08	2150
3D DWT + CNN	62.05	2030	53.67	2030	43.68	2030

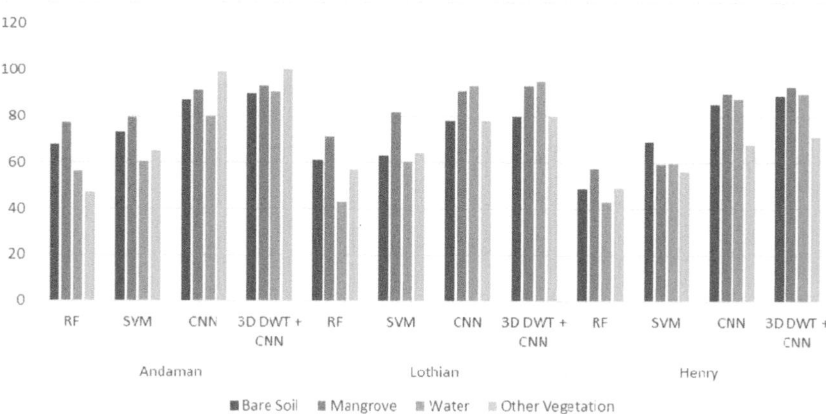

FIGURE 8.8   Overall accuracy of machine learning classifiers tested on different regions.

image-based classification (Table 8.1; Figure 8.8). It is observed that CNN with 3D DWT-based feature extraction helps the CNN model generalize better and gives better overall accuracy than other models. Table 8.2 displays the classification speed of different machine learning models.

The deep learning algorithm was found to operate efficiently for all regions of the data set. However, specific preprocessing methods and fusion of feature extraction methods can be used to improve the accuracy in hyperspectral images. The accuracy and precision of the model were evaluated using the collected hyperspectral image data set, and the results showed that the algorithm achieved high accuracy and precision.

9 Ecodynamic Modeling

9.1 ECODYNAMIC MODELING

The ecodynamic model is a model used to simulate the interactions and dynamics of ecological systems. It is a mathematical model that describes the interactions between different components of an ecosystem, such as the flow of energy and matter, population dynamics, and the effects of disturbances and disturbances on the ecosystem. The ecodynamic model can study a wide range of ecological systems, from small-scale ecosystems like wetlands and forests to large-scale systems like global climate and ocean systems.

One example of applying the ecodynamic model is in the study of wetland ecosystems. Wetlands are important habitats for a wide range of plant and animal species, and they also provide essential ecosystem services, such as water purification, flood control, and carbon sequestration. The ecodynamic model can be used to study the interactions between different components of the wetland ecosystem, such as the flow of water, nutrients, and energy, the dynamics of plant and animal populations, and the effects of human activities on the wetland.

Another example of the application of the ecodynamic model is in the study of global climate change. The model can be used to study the interactions between different components of the Earth's climate system, such as the atmosphere, oceans, and land surface, and the effects of changes in greenhouse gas concentrations on the climate. The ecodynamic model can also be used to study the impacts of climate change on different ecosystems and the services they provide.

In terms of implementation, the ecodynamic model can be implemented in various ways, depending on the specific ecosystem and research question being studied. It can be implemented as a simple mathematical or more complex model that incorporates detailed information about the ecosystem and its components. The model can also be implemented using a variety of programming languages and software, such as R, MATLAB, or GIS software.

This chapter analyzes the interactions among mangrove species through successfully decoding information sets that remain latent in their hyperspectral images.

DOI: 10.1201/9781003432623-9

9.2 THE ECODYNAMIC MODEL

The ecodynamic model assumes that mangrove species spatially adjacent to each other compete within themselves other to survive within limited resources (Obade et al., 2009). This is referred to as the 'survival capacity (SC)' of a population, which is dependent upon its rate of reproduction, rate of mortality, and rate of growth over a period of time.

The survival capacity 'K_i' of species 'i' is calculated as

$$K_i = \frac{\left(a_i - \mu_i\right) * p_i}{a_i}$$

$$K_i = \frac{r_i * p_i}{a_i}$$

where 'a_i' is the rate of reproduction, 'μ_i' the rate of mortality, 'r_i' the growth rate, and 'p_i' the fractional abundance of species 'i' in a pixel area over the years 2011 and 2014. 'r_i' and 'μ_i' have been calculated on the basis of increase and decrease in species population size (difference in fractional abundance occupied by species 'i') with time, respectively. The growth of species over a period of time is assumed to have a minimum mortality rate (0.0005). Mortality bears the assumption of having a minimum growth rate (0.0005). 'a_i' is the summation of 'r_i' and 'μ_i'.

There is another important parameter called the competition factor (CF), which is used to predict species response in competition with various species is. CF is the spatiotemporal probability of occurrence of species – imposed on species 'i' by species 'j' and vice versa. The CF is expressed as each species 'importance value' and divided by 100.

$$\beta_{ij} = \frac{IV_{ij}}{3 * 100}$$

There is another factor called the 'importance value (IV)' of a species, which is an index depicting the relative dominance of a species in the mangrove community amongst other species. This is expressed as

$$RDO_i = \frac{A_i}{\sum_{i=1}^{m} A} * 100$$

where 'RDO_i' is relative dominance, 'A_i' is the fractional abundance area occupied by species 'i' and 'm' is the total number of species present within the pixel area.

Depending upon the competition factor and survival capacity of species, the 'competition coefficient' (CC) is calculated to analyze the interactions between various species based on their reproduction rate. If the competition coefficient between two species ('i' and 'j') is a_{ij} and the reproduction rate of species is f_i, a_{ij} is expressed as

$$a_{ij} = \frac{\left(2 * \beta_{ij} * k\right) + f_i}{f_i}$$

9.2.1 Ecodynamic Model and Time-Series Analyses of Hyperspectral Data

The ecodynamic model, discussed earlier, is applied on a time-series hyperspectral image data of Henry Island (the study area). The images were acquired in two periods: 27 May 2011 and 23 November 2014. The automated target detection algorithm, N-FINDR, is applied on both images to identify pure endmembers (mangrove species) in the densely forested stretches. A spectral unmixing model is applied, with the identified endmembers as input, to obtain the mangrove species abundance maps and change envisaged over the 3 years (Figures 9.1 and 9.2). The fractional abundance values of each species derived from FCLSU (fully constrained linear spectral unmixing) at each pixel area (30 × 30 sq. m. area) of the hyperspectral image represent its extent of occurrence within that area. The survival capacity of the species over the pixel area is calculated to predict its dominance over others. The survival capacity values, however, vary with time depending upon the type of species, competition, and the physical environment around them.

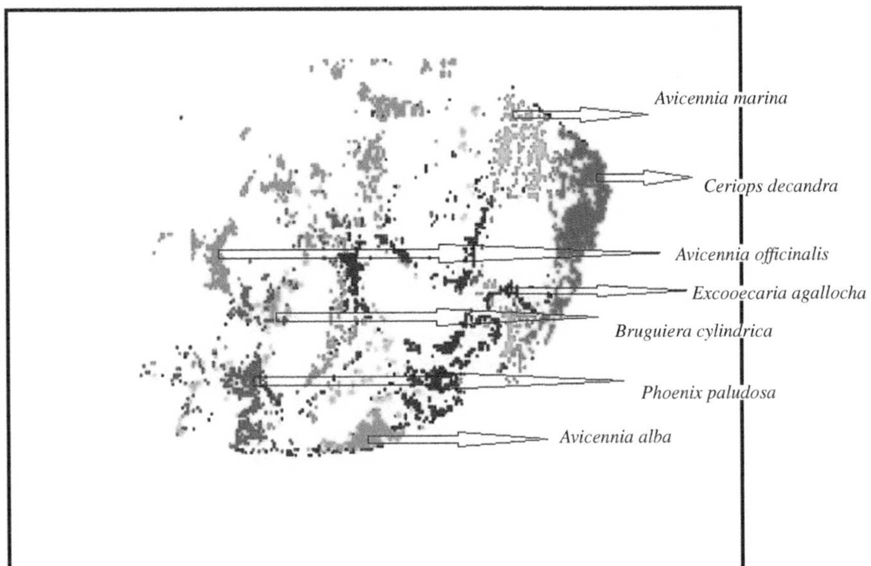

FIGURE 9.1 Map showing mangrove species distribution using the FCLSU algorithm of 2011.

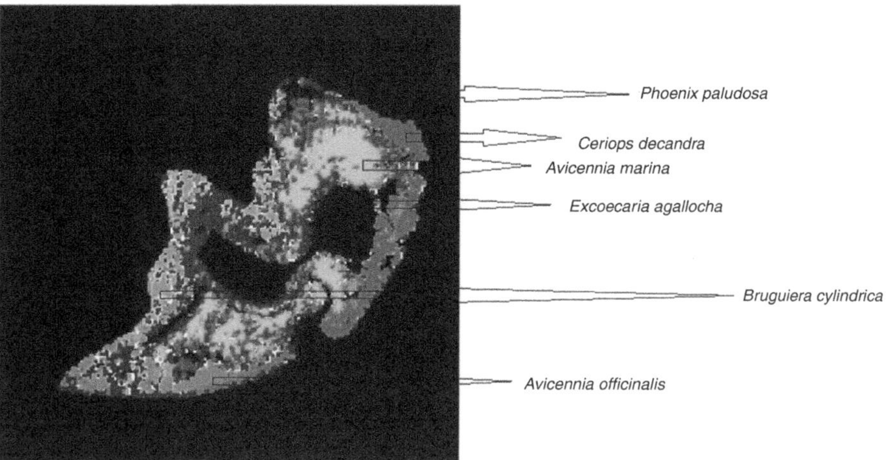

FIGURE 9.2 Map showing mangrove species distribution using the FCLSU algorithm of 2014.

9.2.2 Application of the Ecodynamic Model in Change Detection of Coastal Endmembers

The increase or decrease in abundance of mangrove species in different parts of the study area during 2011 and 2014 is shown in Table 9.1. The time-series accuracy assessment of fractional abundances in the studied locations has been done by calculating the RMSE values (Table 9.1). Though the coordinate locations selected for this study exhibit mixed patch mangroves, the abundance estimation is only calculated based on the singular (linear) interaction portion. The higher-order interactions were ignored to reduce the computational complexity in further calculations. The same was also taken care of during ground truthing.

It is observed from Table 9.1 that, in the 3-year study period at the given location of 21.57702°N; 88.27503°E, the abundance of *Excoecaria agallocha* and *Avicennia alba* have increased slightly, while the abundance of *Avicennia officinalis* and *Bruguiera cylindrica* has decreased within 3 years. The abundance of *Excoecaria agallocha*, *Ceriops decandra*, and *Phoenix paludosa* have decreased at location 21.57308°N; 88.29685°E, whereas *Avicennia officinalis* has been observed to increase from 2011 to 2014.

9.2.3 Application of Ecodynamic Model in Competitive Growth Rate of Coastal Endmembers

In this section, the ecodynamic model has been applied to deduce the competitive growth rate of mangrove species. The reproduction rate and survival capacity of individual mangrove species have been deduced from abundance estimates of endmembers in the sub-pixels. The growth rate has been calculated from changes noted in the hyperspectral image scene of 2011 and 2014. It is assumed that the

TABLE 9.1
Increase or Decrease in Abundance of Mangrove Species in Different Parts of the Study Area during 2011 and 2014

Name of Species	Geographic Location (2014 Image)	Fractional Abundance	Ground Truth	Geographic Location (2011 Image)	Fractional Abundance	Ground Truth
Excoecaria agallocha		0.4045	0.4		0.2264	0.24
Avicennia officinalis	21.57702°N;	0.0529	0.05	21.57702°N;	0.0800	0.08
Avicennia alba	88.27503°E	0.0495	0.04	88.27503°E	0.0404	0.04
Bruguiera cylindrica		0.0343	0.03		0.0467	0.05
Root Mean Square Error			0.0200		0.0188	
Excoecaria agallocha		0.1393	0.14		0.0921	0.10
Ceriops decandra	21.57308°N;	0.1681	0.17	21.57308°N;	0.1279	0.13
Phoenix paludosa	88.29685°E	0.1649	0.16	88.29685°E	0.0822	0.08
Avicennia officinalis		0.0724	0.07		0.1292	0.13
Root Mean Square Error			0.0262		0.0147	

TABLE 9.2
Growth Rate, Mortality Rate, and Reproduction Rate of Mangrove Species

Geographic Location	Name of Species	Growth Rate	Mortality Rate	Reproduction Rate
21.577020N; 88.275030E	*Excoecaria agallocha*	0.01978	0.0005*	0.02028
	Avicennia officinalis	0.0005#	0.0090	0.0095
	Avicennia alba	0.0030	0.0005*	0.0035
	Bruguiera cylindrica	0.0005#	0.0041	0.0046
21.573080N; 88.296850E	*Excoecaria agallocha*	0.0157	0.0005*	0.0162
	Ceriops decandra	0.0134	0.0005*	0.0139
	Phoenix paludosa	0.0275	0.0005*	0.028
21.5769330N; 88.27520E	*Excoecaria agallocha*	0.0526	0.0005*	0.0531
	Avicennia alba	0.0005#	0.0081	0.0086
	Bruguiera cylindrica	0.0005#	0.0039	0.0044
	Avicennia officinalis	0.0531	0.0005*	0.0536

Note: *Minimum Mortality Rate: 0.0005; #Minimum Growth Rate: 0.0005

mortality rate of mangrove species does not change with time. The computed growth rate, mortality rate, and reproduction rate of select locations are given in Table 9.2.

9.2.4 APPLICATION OF THE ECODYNAMIC MODEL IN COMPETITIVE SURVIVAL CAPACITY OF COASTAL ENDMEMBERS

Equilibrium analysis of the mangrove ecosystem has been expressed under varying survival capacities ('K_i' and 'K_j') and interaction coefficients ('a_{ij}' and 'a_{ji}'). Importance value (IV) is another parameter that indicates the dominance of a particular species over another. The higher the IV value of a particular species, the more its domination in the area. The IV and SC of mangrove species of specific locations of the study area are given in Table 9.3.

Analysis of Table 9.3 indicates that *Excoecaria agallocha* is the most dominant and important species depicted on 30m × 30m pixel area at geographical location 21.577020N; 88.275030E. *Avicennia officinalis* is the next dominant species in that region followed by *Avicennia alba* and *Bruguiera cylindrica*. A similar situation persists at coordinates 21.5769330N; 88.27520E.

At coordinate location, 21.573080N; 88.296850E, *Ceriops decandra* is the dominant species, followed by *Excoecaria agallocha* and *Phoenix paludosa*. The matrix representing competition coefficients between different species at location 21.5769330N; 88.27520E is displayed in Tables 9.4–9.6.

On further analysis of survival capacity values and coexistence inequalities (Table 9.7), it is evident that certain degrees of disturbances produce changes in equilibrium status.

TABLE 9.3
Survival Capacity and Importance Values of Mixed Mangrove Species

Geographic Location	Name of Species	Survival Capacity (k)	Importance Value (IV)
21.577020N;	*Excoecaria agallocha*	0.3945	74.7413
88.275030E	*Avicennia officinalis*	0.0028	9.7746
	Avicennia alba	0.0424	9.1463
	Bruguiera cylindrica	0.0037	6.3378
21.573080N;	*Excoecaria agallocha*	0.1350	29.4940
88.296850E	*Ceriops decandra*	0.1621	35.5918
	Phoenix paludosa	0.1620	34.9142
21.5769330N;	*Excoecaria agallocha*	0.3924	72.5857
88.27520E	*Avicennia alba*	0.0020	6.1939
	Bruguiera cylindrica	0.0030	4.7645
	Avicennia officinalis	0.0890	16.4559

TABLE 9.4
Competition Coefficients (of Mangrove Species at Location 21.577020N; 88.275030E)

	Excoecaria agallocha	*Avicennia officinalis*	*Bruguiera cylindrical*	*Avicennia alba*
Excoecaria agallocha	0	1.0008	1.0297	1.0014
Avicennia officinalis	1.0448	0	1.2269	1.0105
Bruguiera cylindrica	2.9812	1.2591	0	1.0112
Avicennia alba	1.1912	1.0250	1.0234	0

TABLE 9.5
Competition Coefficients (of Mangrove Species at Location 21.5769330N; 88.27520E)

	Excoecaria agallocha	*Avicennia alba*	*Bruguiera cylindrical*	*Avicennia officinalis*
Excoecaria agallocha	0	1.0004	1.0009	1.0075
Avicennia alba	1.0536	0	1.0103	1.0882
Bruguiera cylindrica	1.2046	1.0175	0	1.1147
Avicennia officinalis	1.1464	1.0125	1.0096	0

TABLE 9.6
Competition Coefficients (of Mangrove Species at Location 21.573080N; 88.296850E)

	Excoecaria agallocha	*Ceriops decandra*	*Phoenix paludosa*
Excoecaria agallocha	0	1.2814	1.1369
Ceriops decandra	1.1932	0	1.1135
Phoenix paludosa	1.0977	1.1179	0

TABLE 9.7
Inequalities Used to Interpret State of Equilibrium of Mangrove Species

Inequalities	$k_i > (k_j / a_{ij})$	$k_i < (k_j / a_{ij})$
$k_i > (k_j / a_{ij})$	Unstable State	Species j wins
$k_i < (k_j / a_{ij})$	Species i wins	Stable State

Note: k stands for survival capacity and, a_{ij} and a_{ji} are the competition coefficients.

9.2.5 Application of the Ecodynamic Model in Pair-wise Competition of Coastal Endmembers

Combination of species interacting with each other may exhibit a state of equilibrium or disequilibrium due to competition amongst themselves or other external influences. Inequalities have been used to exhibit stability or instability due to pair-wise interactions between mangrove species have been used and displayed in Table 9.1.

At geographic location 21.577020 N; 88.275030 E in Table 9.8, it is observed that *Excoecaria agallocha* exhibits the most stable state when under competition with *Avicennia officinalis*, *Avicennia alba*, and *Bruguiera cylindrica*. *Avicennia officinalis* is the next dominant species in the pixel coordinate, followed by *Avicennia alba*.

At 21.573080 N; 88.296850 E, *Excoecaria agallocha* and *Ceriops decandra* have the same strength to compete with each other resulting in instability where either species can dominate over each other. However, competition between *Excoecaria agallocha* and *Phoenix paludosa* indicates that *Phoenix paludosa* is the more robust species amongst them.

At 21.5769330 N; 88.27520 E, the area is dominated by *Excoecaria agallocha* followed by *Avicennia officinalis*, *Avicennia alba*, and *Bruguiera cylindrica*. It is thus observed from time-series data of 2011 and 2014 that *Excoecaria agallocha*, *Avicennia officinalis*, and *Avicennia alba* have better success rates of survival in competition with other species in the study area.

TABLE 9.8
Pair-wise Competition amongst Species and their Outcomes

Geographic Location	Competing Species	Species 1 Wins	Species 2 Wins	Unstable	Stable
21.577020N; 88.275030E	*Excoecaria agallocha* and *Avicennia officinalis*	Yes			
	Excoecaria agallocha and *Avicennia alba*	Yes			
	Excoecaria agallocha and *Bruguiera cylindrica*	Yes			
	Avicennia officinalis and *Avicennia alba*		Yes		
	Avicennia officinalis and *Bruguiera cylindrica*		Yes		
	Avicennia alba and *Bruguiera cylindrica*	Yes			
21.573080N; 88.296850E	*Excoecaria agallocha* and *Ceriops decandra*			Yes	
	Excoecaria agallocha and *Phoenix paludosa*		Yes		
	Ceriops decandra and *Phoenix paludosa*			Yes	
21.5769330N; 88.27520E	*Excoecaria agallocha* and *Avicennia alba*	Yes			
	Excoecaria agallocha and *Bruguiera cylindrica*	Yes			
	Excoecaria agallocha and *Avicennia officinalis*	Yes			
	Avicennia alba and *Bruguiera cylindrica*	Yes			
	Avicennia alba and *Avicennia officinalis*		Yes		
	Bruguiera cylindrica and *Avicennia officinalis*		Yes		

The Lotka–Volterra Competition Model

The Lotka–Volterra competition model (Bomze, 1983; 1995; Sprott et al., 2005) is used to predict the outcome of competitive interactions of species, and whether the system exhibits a state of equilibrium or disequilibrium (Goel et al., 2003). The differential equations of the competitive Lotka–Volterra model are used for equilibrium analysis of the dynamic mangrove ecosystem for the species competing for a shared resource. This model has been generalized to the 'n' number of species competing against each other and calculates the growth rate of a species in competition with the remaining '$n - 1$' species within the area.

$$\frac{dA_i}{dt} = r_i A_i \left(1 - \sum_{j=1}^{N} a_{ij} A_j \right)$$

$\dfrac{dA_i}{dt}$ is the growth rate of species i with respect to time.

r_i is the intrinsic growth rate of a species i.

A_i is the fractional abundance of a species i in a particular pixel location.

a_{ij} is the competition coefficient of species i with respect to another species.

A_j is the fractional abundances of the other species except for the ith species present in the same pixel area.

N is the total number of species present in a particular pixel location.

Here, the value of 'a_{ii}' is self-interacting terms, considered to be 0 as we consider interspecies interaction rather than intra-species interaction. The species are considered to have logistic dynamics in which the population increases exponentially.

9.3 APPLICATION OF LOTKA–VOLTERRA MODEL IN COMPETITIVE GROWTH RATE OF SPECIES

The application of the Lotka–Volterra model at geographic location 21.577020 N; 88.275030 E registers *Excoecaria agallocha* as species with the fastest growth rate when compared with *Avicennia alba*, *Avicennia officinalis*, and *Bruguiera cylindrica*. The growth rate of *Excoecaria agallocha* also marks complete dominance, followed by *Avicennia officinalis* (Table 9.9).

Location 21.573080 N; 88.296850 E is marked by the maximum growth rate of *Ceriops decandra* in relation to *Excoecaria agallocha* and *Phoenix paludosa*. *Phoenix paludosa*, compared with *Excoecaria agallocha* shows a higher growth rate at the same location.

At location 21.5769330 N; 88.27520 E also, *Excoecaria agallocha* is the dominant species in competition with *Avicennia alba*, *Bruguiera cylindrica*, and *Avicennia officinalis*. However, *Avicennia alba* and *Bruguiera cylindrica* do not show any increase in growth within the study period. *Avicennia alba* and *Bruguiera cylindrica* have emerged as inferior competitors. *Excoecaria agallocha* emerges as the superior competitor amongst all other mangrove species within the area.

TABLE 9.9
Growth Rate of Species in Competition with Lotka–Volterra Model

Geographic Location	Name of Species	Growth Rate
21.57702,88.27503	*Excoecaria agallocha*	.0012
	Avicennia officinalis	.0001
	Avicennia alba	.0000
	Bruguiera cylindrica	.0000
21.57308,88.29685	*Excoecaria agallocha*	.0004
	Ceriops decandra	.0042
	Phoenix paludosa	.0029
21.576933,88.2752	*Excoecaria agallocha*	.0050
	Avicennia alba	.0000
	Bruguiera cylindrica	.0000
	Avicennia officinalis	.0013

The Lotka–Volterra model has uniquely made use of hyperspectral data in deriving the intra- and inter-competitions and connectivities between mangrove species. The model has highlighted the dominant mangrove species of an area, which may henceforth be used to predict the spatio-temporal pattern of survival of species. The study has successfully extracted relevant features from hyperspectral imagery to estimate the fractional abundances of target objects as inputs to derive the relative dominance and importance values of individual species and their survival capacities. The Lotka–Volterra model observes that *Excoecaria agallocha, Avicennia officinalis,*and *Ceriops decandra* are the most dominant species, followed by *Phoenix paludosa* and *Bruguiera cylindrica* in the study area. Further, the dominance of *Excoecaria agallocha* and *Ceriops decandra* in the same area indicates that their same strength to compete with each other for survival, and this leads to instability of the ecosystem at those places.

The findings of the Lotka–Volterra model will help in formulation ofa forest plan for the restoration, regeneration, conservation, and management of mangrove plants in the Sunderban Biosphere Reserve. The procedure adopted can be used as a paradigm exemplar to infer ecocompetitiveness and ecodynamics of other forest biomes as well.

9.4 CHALLENGES OF HYPERSPECTRAL DATA FOR MAPPING AND CLASSIFICATION – GAPS AND SOLUTIONS

Hyperspectral data, also known as imaging spectroscopy data, is a powerful tool for mapping and classification, but it also presents several challenges. Some of the main challenges include the following:

1. High dimensionality: hyperspectral data has a high number of spectral bands, making it difficult to process and analyze. This can lead to problems such as 'the curse of dimensionality,' where the number of features exceeds the

number of samples, and 'spectral mixing,' where different materials have similar spectral signatures.

2. Noise and atmospheric effects: hyperspectral data is often affected by noise and atmospheric effects, such as water vapor and atmospheric gases, making it challenging to extract useful information.

3. Limited availability of ground truth data: hyperspectral data is often used for mapping and classification, but obtaining accurate ground truth data for validation can be challenging.

4. Computational complexity: processing and analyzing hyperspectral data can be computationally intensive and require specialized software and hardware.

Solutions to these challenges include the following:

1. Dimensionality reduction techniques: techniques such as principal component analysis (PCA), independent component analysis (ICA), and linear discriminant analysis (LDA) can be used to reduce the dimensionality of the data and improve the performance of classification algorithms.

2. Noise reduction and atmospheric correction: techniques such as atmospheric correction, noise reduction, and image enhancement can be used to remove noise and correct for atmospheric effects.

3. Collecting ground truth data: ground truth data can be collected using a variety of methods, such as field measurements, ground-based remote sensing, and lab measurements.

4. Machine learning and deep learning methods: these methods can improve the accuracy of mapping and classification by training algorithms on large amounts of data and allowing the algorithm to learn the relationships between the spectral bands and the classes of interest.

Overall, while hyperspectral data has its own set of challenges, it can be overcome by using the proper techniques and tools. By addressing these challenges, the resulting maps and classifications will be more accurate, providing more useful information for different fields such as environmental monitoring, land-use management, and even for mineral exploration.

Bibliography

Altmann, Y., Dobigeon, N., & Tourneret, J.Y. (2011). Bilinear models for nonlinear unmixing of hyperspectral images. *IEEE Transactions on Geoscience and Remote Sensing*, 49(11), 4153–4162. DOI: 10.1109/TGRS.2010.2098414

Aschbacher, J., Tiangco, P., Giri, C.P., Ofren, R.S., Paudyal, D.R., & Ang, Y.K. (1995).Comparison of different sensors and analysis techniques for tropical mangrove forest mapping. In: *Proceedings of the International Conference* IGARSS, 2109–2111.

Asner, G.P. et al., (2000). Impact of tissue, canopy, and landscape factors on the hyperspectral reflectance variability of arid ecosystems. *Remote Sensing of Environment*, 74(1), 69–84. https://doi.org/10.1016/S0034-4257(00)00124-3

Bali, N., & Mohammad-Djafari, A. (2008). Bayesian approach with hidden Markov modelling and mean field approximation for hyperspectral data analysis. *IEEE Transactions on Image Processing*, 17(2), 217–225.

Beck, R. (2003). EO-1 User Guide, Version 2.

Benlin, X., Fangfang, L., Xingliang, M., & Huazhong, J. (2008). Study on independent component analysis application in classification and change detection of multispectral images. *International Archives Photogrammetry, Remote Sensing and Spatial Information Sciences*, XXXVII(Part B7), 871–875.

Bioucas-Dias, J. M., Plaza, A., Dobigeon, N., Parente, M., Du, Q., Gader, P., and Chanussot, J. (2012). Hyperspectral unmixing overview: Geometrical, statistical, and sparse regression-based approaches. *IEEE Journal of Selected Topics in Applied Earth Observations and Remote Sensing*, 5(2), 354–379.

Bomze, I.M. (1983). Lotka–Volterra equation and replicator dynamics: A two dimensional classification. *Biological Cybernetics*, 48, 201–211.

Bomze, I.M. (1995). Lotka–Volterra equation and replicator dynamics: New issues in classification. *Biological Cybernetics*, 72, 447–453.

Borel, C.C., & Gerstl, S.A. (1994). Nonlinear spectral mixing models for vegetative and soils surface, *Remote Sensing of the Environment*, 47(2), 403–416.

Broadwater, J., & Banerjee, A. (2009). A comparison of kernel functions for intimate mixture models. In: *Proceedings of Workshop on Hyperspectral Image and Signal Processing, Evolution in Remote Sensing*, Grenoble, France, 1–4.

Broadwater, J., Chellappa, R., Banerjee, A., & Burlina, P. (2007). Kernel fully constrained least squares abundance estimates. In: *Proceedings of IEEE International Geoscience and Remote Sensing Symposium*, Barcelona, Spain, 4041–4044.

Celik, T. (2009). Unsupervised change detection in satellite images using principal component analysis and means clustering. *IEEE Geoscience and Remote Sensing Letters*, 6(4), 772–776.

Chakrabarti, S. (1995). Late quaternary stratigraphy and an upper Acheulian occupation site in the Kalubhar valley, Saurashtra. *Memoirs-Geological Society of India*, 277–281.

Chakrabarti, S., & Bhattacharya, H.N. (2013). Inferring the hydro-geochemistry of fluoride contamination in Bankura district, West Bengal: A case study. *Journal of the Geological Society of India*, 82, 379–391.

Chakrabarti, S., & Sen, R. (2020). Unravelling the nature of natural inorganic nanoparticles in the Subarnarekha river, Eastern India. *Journal of the Geological Society of India*, 95(3), 327–328. DOI: 10.1007/s12594-020-1436-x

Chakrabarti, S., & Sen, R. (2021). Insights into differential biomining traits of Indian copper sulphides. *Current Science*, 120(12), 1812.

Chakravortty, S. (2011). Hyperspectral data analysis for mangrove species discrimination: A novel approach. *National Journal on Chembiosis*, 2(1), 8–12.

Chakravortty, S. (2013). Analysis of end member detection and subpixel classification algorithms on hyperspectral imagery for tropical mangrove species discrimination in the Sunderbans Delta, India. *Journal of Applied Remote Sensing*, 7(1), 073523.

Chakravortty, S., & Bhondekar, A. (2018). Spatial and spectral quality assessment of fused hyperspectral and multispectral data. *Lecture Notes in Computational Vision and Biomechanics (LNCVB): Biologically Rationalized Computing Techniques For Image Processing Applications*, vol. 25, 133–158.

Chakravortty, S., & Chakrabarti, S. (2012). Quality enhancement of hyperspectral image data through atmospheric correction: A case study of Henry and Lothian Islands of Sunderban biosphere reserve, West Bengal. *International Journal on Applied Bioengineering*, 6(2), 490–501.

Chakravortty, S., & Chakrabarti, S. (2017). Design and development of higher order spectral unmixing model for Mangrove species discrimination. *Proceedings of the National Academy of Sciences, India Section A: Physical Sciences*, 87, 557–566.

Chakravortty, S., & Choudhury, A.S. (2012). Application of unsupervised end member detection algorithms for spectral unmixing of hyperspectral data for mangrove species discrimination. *2012 International Conference on Communications, Devices and Intelligent Systems (CODIS)*, IEEE, 81–84.

Chakravortty, S., & Choudhury, A.S. (2013). Determining spatial location of sub pixels in hyperspectral data for mangrove species identification. *2013 International Conference on Signal Processing, Image Processing & Pattern Recognition*, IEEE, 39–43.

Chakravortty, S., & Das, S. (2017). Integration of high spectral and high spatial resolution image data for accurate target detection. *Computational Intelligence, Communications, and Business Analytics: First International Conference*, CICBA 2017, Kolkata, India, March 24–25, 2017, Revised Selected Papers, Part I, Springer, 179–192.

Chakravortty, S., & Ghosh, S. (2014). Fusion of high spatial and high spectral data for quality enhancement of remotely sensed images. *Proceedings of the International Conference on Computing, Communication & Manufacturing* 2014.

Chakravortty, S., & Ghosh, D. (2018a). Automatic identification of saline blanks and pattern of related mangrove species on hyperspectral imagery. *2018 4th International Conference on Recent Advances in Information Technology (RAIT)*, IEEE, 1–6.

Chakravortty, S., & Ghosh, D. (2018b). Development of a model for detection of saline blanks amongst mangrove species on hyperspectral image data. *Current Science*, 115(3), 541–548.

Chakravortty, S., Ghosh, D., & Sinha, D. (2018). A dynamic model to recognize changes in mangrove species in Sunderban delta using hyperspectral image analysis. *Progress in Intelligent Computing Techniques: Theory, Practice, and Applications: Proceedings of ICACNI 2016*, vol.1, Springer, 59–67.

Chakravortty, S., Li, J., & Plaza, A. (2017). A technique for subpixel analysis of dynamic mangrove ecosystems with time-series hyperspectral image data. *IEEE Journal of Selected Topics in Applied Earth Observations and Remote Sensing*, 11(4), 1244–1252.

Chakravortty, S., & Shah, E. (2013). Application of non-linear spectral unmixing on hyperspectral data for species level classification of mangroves. *2013 International Conference on Communication and Signal Processing*, IEEE, 1123–1127.

Chakravortty, S., Shah, E., & Chowdhury, A.S. (2014). Application of spectral unmixing algorithm on hyperspectral data for mangrove species classification. *Applied Algorithms: First*

International Conference, ICAA 2014, Kolkata, India, January 13-15, 2014. Proceedings 1, Springer, 223–236.

Chakravortty, S., & Sinha, D. (2014). Performance of pure pixel extraction algorithms on hyperspectral data for species level classification of mangroves. *2014 Fourth International Conference of Emerging Applications of Information Technology*, IEEE, 209–214.

Chakravortty, S., & Sinha, D. (2015). Analysis of multiple scattering of radiation amongst end members in a mixed pixel of hyperspectral data for identification of mangrove species in a mixed stand. *Journal of the Indian Society of Remote Sensing,* 43, 559–569.

Chakravortty, S., & Sinha, D. (2016). Development of higher-order model for non-linear interactions in hyperspectral data of mangrove forests. *Current Science,* 111(6),1055–1062.

Chakravortty, S., Sinha, D., & Bhondekar, A. (2015). Assessment of urbanization of an area with hyperspectral image data. *Proceedings of the 3rd International Conference on Frontiers of Intelligent Computing: Theory and Applications (FICTA)* 2014: vol.2, Springer, 315–322.

Chakravortty, S., & Subramaniam, P. (2014). Fusion of hyperspectral and multispectral image data for enhancement of spectral and spatial resolution. *International Archives of the Photogrammetry, Remote Sensing & Spatial Information Sciences,* 40, 1099–1103.

Chang, Yuqing, Petvipusit, Kurt R., & Deepak, D. (2015). Multi-objective optimization coupled with dimension-wise polynomial-based approach in smart well placement under model uncertainty. *SPE Reservoir Simulation Symposium*, Houston, Texas, USA, February 2015.

Chen, F., & Zhang, Y. (2013). Sparse hyperspectral unmixing based on constrained lp - l2 optimization. *IEEE Geoscience and Remote Sensing Letters,* 10(5), 1142–1146.

Chowdhury, A., & Maiti, S.K. (2014). Mangrove reforestation through participation of vulnerable population: Engineering a sustainable management solution for resource conservation. *International Journal of Environmental Research and Development,* 4(1), 1–7.

Cochrane, M.A. (2000). Using vegetation reflectance variability for species level classification of hyperspectral data. *International Journal of Remote Sensing,* 21(10), 2075–2087. https://doi.org/10.1080/01431160050021303

Curran, P.J. (1989). Remote sensing of foliar chemistry. *Remote Sensing of Environment,* 30(3), 271–278. http://dx.doi.org/10.1016/0034-4257(89)90069-2

Curtis, J.T. (Ed.) (1959). *The Vegetation of Winsconsin. An Ordination of Plant Communities.* Madison, WI, USA, Univ. of Wisconsin Press.

Das, S., & Chakravortty, S. (2020). Identification of interfaces of mixtures with nonlinear models. *Proceedings of the Global AI Congress,* 2019, 237–249. Springer.

Das, S., & Chakravortty, S. (2021a). A novel local consensus inspired spatial fuzzy method for classification and spectral unmixing of hyperspectral data. *Proceedings of the International Conference on Computing and Communication Systems: I3CS 2020, NEHU*, Shillong, India, Springer, 137–144.

Das, S., & Chakravortty, S. (2021b). Efficient entropy-based spatial fuzzy c-means method for spectral unmixing of hyperspectral image. *Soft Computing,* 25(11), 7379–7397.

Das, S., & Chakravortty, S. (2022a). Spectral-spatial 3D dynamic trimmed median filter for removal of impulse noise in remotely sensed images. *Multimedia Tools and Applications,* 82(11), 15945–15982. https://doi.org/10.1007/s11042-022-13965-y

Das, S., & Chakravortty, S. (2022b). Unsupervised hybrid change detection using geo-spatial spectral classification of time-series remote sensing datasets. In: *Emerging*

Technologies in Data Mining and Information Security: Proceedings of IEMIS 2022, vol.2, Springer, 27–34.

Demuro, M. & Chisholm, L. (2003). Assessment of Hyperion for characterizing mangrove communities. In: *Proceedings of the International Conference the AVIRIS 2003 Workshop*, 18–23.

Dislich, C., Johst, K., & Huth, A. (2010). What enables coexistence in plant communities? Weak versus strong species traits and the role of local processes, *Ecological Modelling*, 221(19), 2227–2236.

Dobigeon N. & Tourneret J.-Y. (2007). Spectral unmixing of hyperspectral images using a hierarchical Bayesian model. *IEEE International Conference on Accoustics, Speech and Signal Processing-ICASS*. 3.

Du, Q., Raksuntorn, N., Younan, N.H., & King, R. (2008). Variants of N-FINDR Algorithm for Endmember Extraction. In: *Image and Signal Processing for Remote Sensing XIV, Proc. of SPIE*, vol. 7109, 71090 G-1.

Eches, O., Dobigeon, N. & Tourneret, J.-Y. (2011). Enhancing hyperspectral image unmixing with spatial correlations. *IEEE Transactions on Geoscience and Remote Sensing*, 49(11), 4239–4247.

Ecology and Climate Justice (2009).www.unnayan.org/

Eismann, M.T., Joseph, M., & Hardie, R.C. (2008). Hyperspectral change detection in the presence of diurnal and seasonal variations. *IEEE Transactions on Geoscience and Remote Sensing*, 46(1), 237–249.

Elvidge, C.D. (1987). Reflectance characteristics of dry plant materials. In: *Proceeding of the International Conference Remote Sensing of Environment*, 721–733.

Elvidge, C.D. (1990). Visible and near infrared reflectance characteristics of dry plant materials. *International Journal of Remote Sensing*, 11, 1775–1795.

Erturk, A., Iordache, M.D., & Plaza, A. (2016). Sparse unmixing based change detection for multi-temporal hyperspectral images. *IEEE Journal of Selected Topics in Applied Earth Observations and Remote Sensing*, 9(2), 708–719.

Erturk, A., Iordache, M.D., & Plaza, A. (2017).Sparse unmixing with dictionary pruning for hyperspectral change detection. *IEEE Journal of Selected Topics Applied Earth Observations and Remote Sensing*, 10(1), 321–330.

Fan, W., Hu, B., Miller, J., & Li, M. (2009). Comparative study between a new nonlinear model and common linear model for analysing laboratory simulated-forest hyperspectral data. *Remote Sensing of Environment*, 30(11), 2951–2962.

Frederick, A.R., & Mosquera, J. (2000). Is space necessary? Interference competition and limits to biodiversity. *Ecology*, 81(11), 3226–3232.

Gao, J. (1999).A comparative study on spatial and spectral resolutions of satellite data in mapping mangrove forests. *International Journal of Remote Sensing*, 20, 2823–2833.

Ghosh, D., & Chakravortty, S. (2020). Change detection of tropical mangrove ecosystem with subpixel classification of time series hyperspectral imagery. *Artificial Intelligence Techniques for Satellite Image Analysis,* 189–211.

Ghosh, D., & Chakravortty, S. (2022a). Generation of high-resolution spectra from multi-spectral imagery using derivative based learning method. *Advanced Techniques for IOT Applications: Proceedings of EAIT*, 2020, 256–267. Springer.

Ghosh, D., & Chakravortty, S. (2022b). Reconstruction of high spectral resolution multispectral image using dictionary-based learning and sparse coding. *Geocarto International*, 37(25), 1–16.

Ghosh, A., Chakrabarti, S., Biswas, K., & Ghosh, U.C. (2014). Agglomerated nanoparticles of hydrous Ce (IV)+ Zr (IV) mixed oxide: Preparation, characterization and physico-chemical aspects on fluoride adsorption. *Applied Surface Science*, 307, 665–676.

Ghosh, A., Chakrabarti, S., Biswas, K., & Ghosh, U.C. (2015). Column performances on fluoride removal by agglomerated Ce (IV)--Zr (IV) mixed oxide nanoparticles packed fixed-beds. *Journal of Environmental Chemical Engineering*, 3(2), 653–661.

Ghosh, A., Chakrabarti, S., & Ghosh, U.C. (2014). Fixed-bed column performance of MN-incorporated iron (III) oxide nanoparticle agglomerates on As (III) removal from the spiked groundwater in lab bench scale. *Chemical Engineering Journal*, 248, 18–26.

Ghosh, D., Chakravortty, S., Plaza, A., & Li, J. (2021). Change prediction and modeling of dynamic mangrove ecosystem using remotely sensed hyperspectral image data. *Journal of Applied Remote Sensing*, 15(4), 042606–042606.

Goel, P.K., Prasher, S.O., Landry, J.A., Patel, R.M., Viau, A.A., & Miller, J.R. (2003). Estimation of crop biophysical parameters through airborne and field hyperspectral remote sensing. *Transactions of the American Society of Agricultural Engineers*, 46, 1235–1246.

Graña, M., & DAnjou, M. (2005). Feature Extraction by Linear Spectral Unmixing. Workshop on Intelligent Information Processing for Remote Sensing, Spain.

Green, E.P., Clark, C.D., & Edwards, A.J. (2000).Image Classification and Habitat Mapping. In: Edwards, A.J. (Ed.), *Remote Sensing Handbook for Tropical Coastal Management*, Paris, UNESCO,141–154.

Green, R.O. et al. (1998). Imaging spectroscopy and the airborne visible/infrared imaging spectrometer (AVIRIS). *Remote Sensing of Environment*, 65(3), 227–248. https://doi.org/10.1016/S0034-4257(98)00064-9

Guilfoyle, K.J., Althouse, M.L., & Chang, C.-I. (2001). A quantitative and comparative analysis of linear and nonlinear spectral mixture models using radial basis function neural networks. *IEEE Geoscience and Remote Sensing Letters,* 39(8), 2314–2318.

Gulati, V., & Pal, P.A. (2014). Survey on various change detection techniques for hyperspectral images. *International Journal of Advances Research in Computer Science and Software Engineering*, 4(8), 852–855.

Halimi, A., Altmann, Y., Dobigeon, N., & Tourneret, J.Y. (2011). Nonlinear unmixing of hyperspectral images using a generalized bilinear model. *IEEE Transactions on Geoscience and Remote Sensing*,49(11).

Han, T., Goodenough, D.G., Dyk, A., & Love, J. (2002).Detection and correction of abnormal pixels in hyperion images, *IGARSS*, 1327–1330.

Hapke, B. (1993). *Theory of reflectance and emittance spectroscopy, topics in remote sensing.* Cambridge, UK: Cambridge University Press.

Heinz, D. C, Chang, C. I. (2001). Fully constrained least squares linear spectral mixture analysis method for material quantification in hyperspectral imagery. *IEEE Transactions on Geoscience and Remote Sensing,* 39(3), 529–545.

Held A., Ticehurst C., Lymburner L. & Williams N. (2003). High resolution mapping of tropical mangrove ecosystems using hyperspectral and radar remote sensing, *International Journal of Remote Sensing*, 24, 2739–2759.

Heylen, R. Parente M. & Gader, P. (2014). A review of nonlinear hyperspectral unmixing methods, *IEEE Journal of Selected Topics in Applied Earth Observations and Remote Sensing*, 7(6), 1844–1868.

Himmelsbach, D.S., Boer, J.D., Akin, D.E., & Barton, E.E. (1988). Solid state carbon-13 NMR, FTIR, and NIR spectroscopic studies of ruminant silage digestion. In: Creaser, C.S., Davies, A.M.C. (Eds.), *Analytical Applications of Spectroscopy*. Royal Society of Chemistry, London, pp. 410–413.

Hirano, A., Madden, M., & Welch, R. (2003). Hyperspectral image data for mapping wetland vegetation. *Wetlands* 23, 436–448.

Hsuang, R., & Chang, C. (2003). Automatic spectral target recognition in hyperspectral imagery. *IEEE Transactions on Aerospace and Electronic Systems*, 39(4).

Huck, A., Guillaume, M., & Blanc-Talon, J. (2010). Minimum dispersion constrained non-negative matrix factorization to unmix hyperspectral data. *IEEE Transactions on Geoscience and Remote Sensing*, 48(6), 2590–2602.

Hussain, M., Chen, D., Cheng, A., Wei, H., & Stanley, D. (2013). Change detection from remotely sensed images: From pixel-based to object-based approaches. *ISPRS Journal of Photogrammetry Remote Sensing*, 80, 91–106.

Jafari, R., & Lewis, M.M. (2012). Arid land characterization with EO-1 hyperion hyperspectral data. *International Journal of Applied Earth Observation and Geoinformation*, 19, 298–307.

Javier, P., Plaza, A., & Martin's, G. (2009). A fast sequential endmember extraction algorithm based on unconstrained linear spectral unmixing. In: Bruzzone, L., Notarnicola, C., Posa, F. (Eds.), *Image and Signal Processing for Remote Sensing XV. Proc. of SPIE*, vol. 7477, 74770L.

Jensen, J.R. (1996). *Introductory Digital Image Processing: A Remote Sensing Perspective.* Pearson, Englewood Cliffs, NJ, USA, Prentice Hall.

Jia, M., Shang, Y., Wang, Z., Song, K., & Ren, C. (2014). Mapping the distribution of mangrove species in the core zone of Mai Po Marshes Nature Reserve, Hong Kong, using hyperspectral data and high-resolution data. *International Journal of Applied Earth Observation and Geoinformation*, 33, 226–231.

John, R., Dattaraja, H.S., Suresh, H.S., & Sukumar, R. (2010). Density dependence in common tree species in a tropical dry forest in Mudumalai, southern India. *Journal of Vegetation Science*, 13(1), 45–56.

Kamal, M., & Phinn, S. (2011). Hyperspectral data for mangrove species mapping: A comparison of pixel-based and object-based approach. *Remote Sensing*, 3(10), 2222–2242.

Kanniah, K., Wai, N., Shin, A., & Rasib, A. (2007). Per-pixel and sub-pixel classifications of high-resolution satellite data for mangrove species mapping. *Applied GIS*,3(8), 1–22.

Kanniah, K. et al. (2007). Per-pixel and sub-pixel classifications of high-resolution satellite data for mangrove species mapping. *Applied GIS*, 3(8), 1–22.

Kanniah, K.D., Ng, S.W., Lau, A.M.S., & Rasib, A.W. (2005). Linear Mixture Modelling applied to Ikonos data for Mangrove Mapping. In: *Proceedings of the 26th Asian Conference on Remote Sensing*, ACRS 2005, 7–11 Nov. 2005, Hanoi, Vietnam.

Keshava, N., & Mustard, J.F. (2002). Spectral unmixing. *IEEE Signal Processing Magazine*, 19(1), 44–57.

Koedsin, W., & Vaiphasa, C. (2013). Discrimination of tropical mangroves at the species level with EO-1 Hyperion data. *Remote Sensing*, 5(7), 3562–3582.

Kruse, F.A. et al., (2000). HyMap: An Australian hyperspectral sensor solving global problem results from USA HyMap data acquisitions. In *Proceedings of the 10th Australasian Remote Sensing and Photogrammetry Conference*, Adelaide, Australia, 18–23.

Kumar, L., Schmidt, K., Dury, S., & Skidmore, A.K. (2001). Imaging spectrometry and vegetation science. In: van de Meer, F., de Jong, S.M. (Eds.), *Imaging Spectrometry*, Dordrecht, Kluwer Academic Press, 111–155.

Kumar, P., & Chakravortty, S. (2021). Generation of sub-pixel-level maps for mixed pixels in hyperspectral image data. *Current Science*, 120(1), 166.

Kumar, V., & Garg, K.D. (2013). Valuable approach for image processing and change detection on synthetic aperture radar data. *International Journal of Current Engineering Technology*, 3(2), 512–516.

Li, N., Huo, H., & Zhao, Y. A. (2013). Spatial clustering method with edge weighting for image segmentation. *IEEE Geoscience and Remote Sensing Letters*, 10(5), 1124–1128.

Licciardi, G., & Del, F. (2011). Pixel unmixing in hyperspectral data by means of neural networks. *IEEE Transactions on Geoscience and Remote Sensing*, 49(11), 4163–4172.

Liu, W., & Wu, E. (2005). Comparison of non-linear mixture models: sub-pixel classification. *Remote Sensing of Environment*, 94, 145–154.

Long, B.G., & Skewes, T.D. (1996). A technique for mapping mangroves with Landsat TM Satellite Data and Geographic Information System. *Estuarine and Coastal Shelf Science*, 43, 373–381.

Luo, W. et al., (2016). A new algorithm for bilinear spectral unmixing of hyperspectral images using particle swarm optimization. *IEEE Journal of Selected Topics Applied Earth Observations and Remote Sensing*, 9(12), 5776–5790.

Miao, L., & Qi, H. (2007). Endmember extraction from highly mixed data using minimum volume constrained nonnegative matrix factorization. *IEEE Transactions on Geoscience and Remote Sensing*, 45(3), 765–777.

Nandy, S., & Kushwaha, S.P.S. (2010). Geospatial modelling of biological richness in Sunderbans, *Journal of ISRS*, 38, 431–440.

Nascimento, J.M.P., & Bioucas-Dias, J.M. (2009). Non-linear mixture model for hyperspectral unmixing. In: Bruzzone, L., Notarnicola, C., and Posa, F. (Eds.), *Proc. SPIE Image and Signal Processing for Remote Sensing* XV, vol. 7477. SPIE: Berlin, Germany. 770I-1–74 770I-8

Nascimento, J.M.P. & Bioucas-Dias, J.M. (2010). Unmixing hyperspectral intimate mixtures. In: Bruzzone, L. (Ed.), *Proc. SPIE Image and Signal Processing Remote Sensing*, XVI, 74830, SPIE: Bellingham, WA, 78300C.

Nayek, S., & Bahuguna, A. (2001). Application of remote sensing data to monitor mangroves and other coastal vegetation of India. *Indian Journal of Marine Science*, 30(4), 195–213.

Obade, P.T. et al., (2009). Impact of anthropogenic disturbance on a mangrove forest assessed by a 1d cellular automaton model using Lotka–Volterratype competition, *International Journal of Design and Nature Ecodynamics*, 3(4), 296–320.

Plaza, A., Martinez, P., Perez, R., &Plaza, J.A. (2004). Quantitative and comparative analysis of endmember extraction algorithms from hyperspectral data. *IEEE Transactions on Geoscience and Remote Sensing*, 42(3), 650–663.

Plaza, J., Martinez, P., Perez, R., & Plaza, A. (2005). Nonlinear neural network mixture models for fractional abundance estimation in AVIRIS hyperspectral images. In: *Proceedings of XIII JPL Airborne Earth Science Workshop*, Pasadena, CA.

Plaza, J., Plaza, A., Martinez, R., & Martinez, P. (2007). *Joint Linear/Nonlinear Spectral Unmixing of Hyperspectral Image*. Data. I-4244-1212-9/07/25.00c2007IEEE

Prasad, P.R.C., Reddy, C.S., Rajasekhar, G., &Dutt, C.B.S. (1992).Mapping and analyzing vegetation types of North Andaman Islands, India. *GIS development > Geospatial Application Papers > Natural Resource Management > Overview*, 3(1).

Ray, S., Paul, G., & Chakrabarti, S. (2023). *Fluoride in Pleistocene Barind Terrace of South Dinajpur, West Bengal: Scope for Nano-Remediation*. Springer.

Roy, A., Chakrabarti, S., & Sen, R. (2013). Self-organized criticality and the Sikkim earthquake (2011) with comparative exemplars. *Current Science,*600–603.

Samanta, K., & Hazra, S. (2012). Landuse/Landcover change study of Jharkhali Island Sundarbans, West Bengal using Remote Sensing and GIS. *International Journal of Geomatics and Geosciences*, 3(2), 299–306.

Saxena, A., Rawat, J.K., & Singh, S.K. (2004). Survey and mapping of mangrove cover using remote sensing—a case study of Sundarbans. www.gisdevelopment.net/application/nrm/coastal/mnm/index.htm

Schmidt, K.S., & Skidmore, A.K. (2003). Spectral discrimination of vegetation types in a coastal wetland. *Remote Sensing of Environment*, 85, 92–108.

Sen, R., & Chakrabarti, S. (2005). Environmental systems collapse: A qualitative critique with an Exemplar. *Journal of Environmental Systems*, 32(4), 335–348.

Sen, R., & Chakrabarti, S. (2007). Environmental resources and their economics for use. *Current Science,*1673–1683.

Sen, R., & Chakrabarti, S. (2009). Biotechnology--applications to environmental remediation in resource exploitation. *Current Science*, 97, 768–775.

Sen, R., & Chakrabarti, S. (2010). Whither imperatives of environmental justice? *Current Science*, 98(4), 476–477.

Sen, R., & Chakrabarti, S. (2011). Why environment is a systems science: A commentary. *International Journal of Environmental Sciences*, 2(2), 1017–1020.

Sen, R., & Chakrabarti, S. (2012a). Nanoscience pursuits in mineral particles and their environmental implications. *International Journal of Environmental Sciences*, 2(4), 2295–2311.

Sen, R., & Chakrabarti, S. (2012b). The arsenic quagmire in Bengal aquifers- the missing links. *Current Science (Bangalore)*, 103(8), 881–882.

Sen, R., & Chakrabarti, S. (2017). Environmental mineralogy and geochemistry-The case of uranium contamination. *Geological Society of India*, 90(1), 124–125.

Sen, R., & Chakrabarti, S. (2019a). An example of systems hierarchy in the aggregation of pyritic framboids in natural aquatic environment. *Geological Society of India*, 93(1), 123–124.

Sen, R., & Chakrabarti, S. (2019b). Disaster management dynamics--An analysis of chaos from the flash flood (2013) in the fragile Himalayan system. *Journal of the Geological Society of India*, 93, 321–330.

Sen, R., Chakrabarti, S., & Chakravortty, S. (2014). Measuring the impacts of land use on water quality influenced by non-point sources. *Current Science*, 1719–1725.

Shakoor, A. (2003). Subpixel Classification of Ground Surface Features. GIS@ Development. pp 1–9.

Shaoqing, Z., & Lu, X. (2008). The comparative study of three methods of remote sensing image change detection, *Int. Archives Photogrammetry, Remote Sensing and Spatial Information Sciences*, 37(B7), 1–4.

Singh, S., & Talwar, R.A. (2014). Comparative study on change vector analysis based change detection techniques, *Sadhan*, 39(6), 1311–1331.

Sirkeci, B., Brady, D., & Burman, J. (2000). Restricted total least squares solutions for hyperspectral imagery. *Proceedings of IEEE International Conference on Accoustics, Speech, and Signal Processing*, 1, 624–627.

Somdatta, C., & Chakrabarti, S. (2011). Pre-processing of hyperspectral data: A case study of Henry and Lothian Islands in Sunderban Region, West Bengal, India. *International Journal of Geomatics and Geosciences*, 2(2), 490.

Somers, B., Tits, L., & Coppin P. (2014). Quantifying non-linear spectral mixing in vegetated areas: Computer simulation model validation and first Results. *IEEE Journal of Selected Topics in Applied Earth Observations and Remote Sensing*, 7(6), 1956–1965.

Sprott, J.C., Wildenberg, J.C., &Azizi, Y. (2005). A simple spatiotemporal chaotic Lotka–Volterra model. *Chaos, Solitons & Fractals*, 26, 1035.

Su, Y., Li, J., Plaza, A., Marinoni, A., Gamba, P., & Chakravortty, S. (2019). DAEN: Deep autoencoder networks for hyperspectral unmixing. *IEEE Transactions on Geoscience and Remote Sensing*, 57(7), 4309–4321.

Tan, K., Jin, X., Plaza, A., Wang, X., Xiao, L., &Du, P. (2016). Automatic change detection in high-resolution remote sensing images by using a multiple classifier system and spectral-spatial features, *IEEE Journal of Selected Topics in Applied Earth Observations and Remote Sensing*, 9(8), 3439–3451.

Tilman, D. (1994). Competition and biodiversity in spatially structured habitats, *Ecology*, 75(1), 2–16.

Tucker, L. (1966). Some mathematical notes on three-mode factor analysis. *Psychometrika*, 31, 279–311

Vaiphasa, C. (2005). Remote sensing techniques for mangrove mapping (thesis), M.S. thesis, Wageningen Univ., Wageningen, The Netherlands.

Vaiphasa, C., & Ongsomwang S. (2005). Tropical mangrove species discrimination using hyperspectral data: A laboratory study, *Estuarine, Coastal and Shelf Science*,65(1–2), 371–379. 10.1016/j.ecss.2005.06.014

Viennois, G. et al., (2016, Aug). Multi-temporal analysis of high-spatial resolution optical satellite imagery for mangrove species mapping in Bali, Indonesia, *IEEE Journal of Selected Topics in Applied Earth Observations and Remote Sensing*, 9(8), 3680–3686.

Wang, L. & Sousa, W.P. (2009). Distinguishing mangrove species with laboratory measurements of hyperspectral leaf reflectance. *International Journal of Remote Sensing*, 30, 1267–1281.

Wang, T., Zhang, H., Lin, H., & Fang, C. (2016). Textural–spectral feature based species classification of mangroves in Mai Po Nature reserve from worldview-3 imagery, *Remote Sensing*, 8(24), 1–15.

Williams, P., & Norris, K. (2001). Near-infrared Technology in the Agricultural and Food Industries. American Association of Cereal Chemists Press, Minnesota, 296 pp.

Winter, M. (1999). Fast autonomous spectral end-member determination in hyperspectral data. In: *Proceedings of the 13th International Conference on Applied Geologic Remote Sensing*, vol. 2, Vancouver, 337–344.

Wu, C., Du, B., &Zhang, L. (2013). A subspace-based change detection method for hyperspectral images, *IEEE Journal of Selected Topics in Applied Earth Observations and Remote Sensing*, 6(2), 815–830.

Wu, X., Li, X., &Zhao, L. (2010). A kernel spatial complexity-based nonlinear unmixing method of hyperspectral imagery. In: Li, K., Jia, L., Sun, X., Fei, M., Irwin, G.W. (Eds.), *LSMS/ICSEE 2010. LNCS*, vol. 6330, Heidelberg, Springer,451–458.

Xu, X., Tong, X., Plaza, A., Zhong, Y., Xie, H., & Zhang, L. (2017). Using linear spectral unmixing for sub-pixel mapping of hyperspectral imagery: A quantitative assessment, *IEEE Journal of Selected Topics Applied Earth Observations and Remote Sensing*, 10(4), 1589–1600.

Yang, C., Everitt, J.H., & Bradford, J.M. (2007). Using multispectral imagery and linear spectral unmixing Techniques for estimating crop yield variability. *Transactions of the ASABE*, 50(2), 667–674. ISSN 0001-2351

Yang, C., Everitt, J.H., & Du, Q. (2010). Applying linear spectral unmixing to airborne hyperspectral imagery for mapping yield variability in grain SORGHUM and cotton fields. *Journal of Applied Remote Sensing*, 4(1), 041887. doi:10.1117/1.3484252.

Yang, C., Everitt, J.H., Fletcher, R. S., Jensen, R.R., & Mausel, P.W. (2009). Evaluating AISA+ hyperspectral imagery for mapping black Mangrove along the South Texas Gulf Coast. *Photogrammetry Engineering and Remote Sensing*, 75, 425–435.

Zare, A. & Ho, K. (2014). Endmember variability in hyperspectral analysis: addressing spectral variability during spectral unmixing. *IEEE Signal Processing Magazine*, 31(1), 95–104.

Zeng, Y., Schaepman, M.E., Wu, B., Clevers, J.G.P.W. & Bregt, A.K. (2007). Forest structural variables retrieval using EO-1 hyperion data in combination with linear spectral unmixing and an inverted geometric-optical model. *Journal of Remote Sensing (China)*, 11(5), 648–658.

Zhi-Gang, L. et al., (2008). Local spatial statistics for remotely sensed image classification of mangrove, *International Archieves of the Photogrammetry, Remote Sensing and Spatial Information Sciences,* XXXVII(Pt.B7), 719.

Zhuang, L., Zhang, B., Gao, L., Li, J., & Plaza, A. (2015). Normal endmember spectral unmixing method for hyperspectral imagery, *IEEE Journal of Selected Topics Applied Earth Observations and Remote Sensing*, 8(6), 2598–2606.

Index